DEVALUED AND DISTRUSTED

DEVALUED AND DISTRUSTED
CAN THE PHARMACEUTICAL INDUSTRY RESTORE ITS BROKEN IMAGE?

JOHN L. LAMATTINA, Ph.D.

WILEY

A JOHN WILEY & SONS, INC., PUBLICATION

All trademarks are property of their respective owners. Lipitor® is a registered trademark of Pfizer Ireland Pharmaceuticals. Chantix®, Viagra®, Zoloft®, and Norvasc® are registered trademarks of Pfizer, Inc.

For general information on our other products and services or for technical support, please contact our Customer Care Department within the United States at (800) 762-2974, outside the United States at (317) 572-3993 or fax (317) 572-4002.

Wiley also publishes its books in a variety of electronic formats. Some content that appears in print may not be available in electronic formats. For more information about Wiley products, visit our web site at www.wiley.com.

Library of Congress Cataloging-in-Publication Data:
LaMattina, John L.
    Devalued and distrusted : can the pharmaceutical industry restore its broken image? / John L. LaMattina, PhD.
        pages cm
    Includes index.
    ISBN 978-1-118-48747-1 (pbk.)
    1. Pharmaceutical industry–Corrupt practices.   I. Title.
    HD9665.5.L36 2013
    338.4'76151–dc23
                        2012032915

10 9 8 7 6 5 4 3 2 1

For Ryer Patrick LaMattina

CONTENTS

ACKNOWLEDGMENTS

I AM particularly indebted to the scientists around the world whose dedication, insights, and acumen produced the important medicines discussed in this book. Many lives have been saved and extended by their efforts. They all deserve our respect and admiration.

I would like to thank a few people for their advice and encouragement during the preparation of this book: David LaMattina and Mary LaMattina for their guidance on key issues; Stephen Lederer and Justin McCarthy for their insightful editorial comments; and Donna Green, who pulled it all together.

J.L.

INTRODUCTION

"**H**ELLO, **MY** name is Rosalyn Menon. I am a producer of *The Dr. Oz Show*. I was wondering if you'd be interested in appearing on an upcoming show."

Rosalyn's phone call took me by surprise. Her interest in me stemmed from the fact that I had written a book a few years earlier called *Drug Truths— Dispelling the Myths About Pharma R&D*. I had written *Drug Truths* after retiring from Pfizer, where I had worked for over 30 years and had been the President of Pfizer's Global R&D Division. *Drug Truths* was written out of the frustration I had over the misperceptions about the drug industry, particularly concerning the lack of understanding that most people have regarding the value that pharmaceutical R&D brings to the health of people around the world. I had a vague familiarity with the show and understood it to be one that tried to educate the public on medical matters.

Rosalyn was now giving me an opportunity to help bring my message to the followers of Dr. Oz. She asked if I'd be willing to appear on the show to talk about issues that I addressed in *Drug Truths*: the safety of new drugs, the accusation that the pharmaceutical company invents diseases, that people are overmedicated, and so on. She told me that Dr. John Abramson, author of *Overdosed America*, would also appear. Abramson is known as a harsh critic of the pharmaceutical industry, and the opportunity to be able to debate him on some of these points appealed to me. So, I agreed to do the show.

About 40 hours later, I was at the NBC studios at Rockefeller Center for the beginning of a unique and challenging experience. I was placed in a Green Room (yes, it's really green!) and was able to watch (on a monitor) preparations for the taping of the show, which would air three weeks later. Parts of what I observed were amusing. The audience, which was already seated, was asked to display various emotions that the show's director would later insert into the segment. On cue, the audience was asked to laugh, express disapproval, or even gasp in horror, thereby dashing my perceptions that the audience reactions seen on talk shows are spontaneous.

My amusement faded pretty quickly as I walked onto the stage of *The Dr. Oz Show*. I was stunned to see the backdrop—a huge banner that read: "The Four Secrets that Drug Companies Don't Want You to Know"! In my naiveté, I never thought to ask Rosalyn for the title of my segment. I quickly realized that the balanced debate

Devalued and Distrusted: Can the Pharmaceutical Industry Restore Its Broken Image?
First Edition. John L. LaMattina.
© 2013 John Wiley & Sons, Inc. Published 2013 by John Wiley & Sons, Inc.

I had expected was going to be tilted. Things got worse with the announcer's opening statement:

> It was seen as a miracle solution. The cure to obesity in a tiny pill. . . . But inside this pill could be a nightmare in disguise, linked to severe liver disease, acute pancreatic damage, and kidney stones.

The drug was orlistat, sold as either Xenical by prescription or Alli over the counter. Orlistat acts by inhibiting fat absorption from the intestine. That sounds pretty good, but here are the consequences: Fat that you eat doesn't get absorbed but rather gets excreted directly, resulting in foul-smelling stools, diarrhea, and "anal leakage." Orlistat can cause modest weight loss, but these toleration issues mean it's a drug many people avoid. But what about the "nightmare" side effects? In 2010, the FDA announced that they had found 13 cases of severe liver damage, one in the United States with Alli and 12 foreign reports with Xenical, over the period of April 1999 to August 2009. It was estimated that over this timeframe, over 40 million people took orlistat. Dr. Oz mentioned this number later in the show, but by then the seeds had already been sown with the audience: Drug companies make unsafe compounds and try to hide the danger of these medicines from consumers. There were two people in the audience who were taking orlistat. Dr. Oz asked them what they thought of this new information, and they expressed shock that they had been prescribed such an unsafe drug. I wonder what they would have thought if Dr. Oz told them that acetaminophen, the active ingredient of Tylenol, is potentially far more harmful to the liver than orlistat?

It was clear from watching the faces of the audience over the next half-hour that they didn't buy my explanations of what pharmaceutical companies do to ensure the safety and efficacy of new medicines. Their view, simply, was that drug companies are out to make money at the expense of the unsuspecting public. This assumption is a major problem facing the pharmaceutical industry today. But the problem runs far deeper than that.

As I walked off the stage after the taping, an angry woman came up to me and said: "The pharmaceutical industry killed my daughter."

I was stunned. I felt as if I had been punched in my gut. I tried to express my sorrow, but she abruptly turned and walked away. This woman suffered an unfathomable loss. I cannot think of anything more tragic than losing a child. Yet, I had no idea what caused her daughter's death. Did her daughter suffer a rare, unforeseen, severe adverse drug reaction? Was the drug improperly administered? How could such a horrible thing happen? Regardless of how this happened, such a loss of life is both dreadful and unintentional. The goal of anyone in pharmaceutical R&D is to alleviate pain and suffering, not cause it. Unfortunately, this woman will never believe that. I fear that many share her views.

This negative view of the pharmaceutical industry is a major problem. But it is not the only one. The industry is also being assailed for the perception that its R&D engines are broken and perhaps unfixable. Furthermore, there seems to be no shortage of "experts" with solutions on how to fix the problem. Unfortunately, many of the proposals are shockingly naïve and without merit. Yet, there are medical needs in diseases like Alzheimer's, diabetes, mental illness, and so on, that are crying out for new drugs and where the pharmaceutical companies can add tremendous value in finding new treatments.

This book critically examines public perceptions and industry realities. Starting with "Dr. Oz's Four Secrets," it addresses the progress the industry has made in improving its abilities to measure both risk and benefits of its new medicines. It then tackles the issue of R&D productivity: What are the reasons for the drop-off and is it temporary or permanent? Next, where should the industry invest its R&D resources? Of the multitude of solutions being offered to fix R&D, what suggestions are of value and how should a company proceed in the difficult business climate that it is now facing? Finally, how can the industry rehabilitate its image?

We are in the midst of a period when medical science has unprecedented opportunities thanks to the knowledge being generated by the Human Genome Project and other profound advances in knowledge. The pharmaceutical industry can play an important role in converting a good deal of this knowledge into new medicines. There isn't one of us who wouldn't benefit from its success. But the pharmaceutical industry is facing an uphill battle in capitalizing on this knowledge. Converting science into medicines is increasingly challenging. Patients, physicians, regulators, and payers need a more accurate appreciation of these challenges plus the value that pharmaceutical R&D adds to society. Without that understanding, the pharmaceutical industry risks being isolated from these groups. That separation could destroy innovation and deprive us all of new and better therapies.

CHAPTER 1

THE FOUR SECRETS THE DRUG COMPANIES DON'T WANT YOU TO KNOW

Pretending to care about our health is often just a part of the drug and other medical industries' overall strategy to increase their sales.

—Dr. John Abramson, *Overdosed America*

A comic strip entitled "Off the Mark" (written by Mark Parisi) depicted the following: The Tin Man comes upon Dorothy and the Cowardly Lion in the woods. At their feet are the remains of the Scarecrow. Dorothy, who, along with the Lion, is eating the last bits of straw, says to the horrified Tin Man: "Dr. Oz said to eat more whole wheat. . . ."

Admittedly, this cartoon takes the influence of Dr. Oz to the extreme. But its premise is not too far from the truth. Dr. Oz has become America's physician. He has the endorsement of Oprah Winfrey and his own daily show (twice a day in some markets!). In 2011, he won two Daytime Emmy awards, one for best informative talk show and the other for best talk-show host. His show on average is watched by over 4 million people each day. He is clearly beloved by his followers. He is better known and, when it comes to dispensing medical advice, he is more influential than the US Surgeon General (Dr. Regina Benjamin at the time of this writing).

But this fame and influence puts an onus on him to be especially accurate in his comments and opinions. When I appeared on Dr. Oz's show with Dr. John Abramson, we debated the value of statin drugs to prevent heart attacks and strokes: Dr. Abramson took the view that statins are unnecessarily prescribed while I defended the fact that statins have saved countless lives. Dr. Abramson said that he often took his patients off statins and I was stunned by this. I turned to Dr. Oz and asked: "Dr. Oz, you're a cardiologist, how do you prescribe statins?" He replied: "I am usually the one taking them off statins," at which point the audience broke into applause.

Dr. Oz then said that he doesn't have a vendetta against statins and that: "There are people whose lives are saved every day with statins." However, I am afraid that

Devalued and Distrusted: Can the Pharmaceutical Industry Restore Its Broken Image?
First Edition. John L. LaMattina.

these later points were lost to both the audience and the millions who watched the show. What many people actually heard from Dr. Oz's pronouncement was that millions of people are needlessly taking statins. This is evident from subsequent summaries of this discussion that now appear on various websites. These summaries highlight the fact that Dr. Oz takes patients off statins and don't mention his points about how statins do save lives.

Like Dr. Oz, I am a big believer in making lifestyles changes, such as watching your diet or exercising, as a way of warding off disease and also preventing the need for medications. But I believe that doctors don't haphazardly prescribe statins to their patients. They prescribe these drugs knowing their patients' health history and medical profiles. When a patient hears someone as prominent as Dr. Oz saying that medicines like statins are overprescribed, many are likely to stop taking these drugs—resulting in potential dire downstream consequences.

Dr. Oz does a great job with certain things such as warning his audience about the perils of too much sugar in their diet. He is a great teacher of good nutrition and the benefits of eating fruits and vegetables. He often provides sound advice and information. But Dr. Oz is in a unique position among daytime hosts. His words are gospel to the American public. Thus, he has an enormous responsibility when it comes to commenting on medicines that members of his own profession prescribe. Unfortunately, there will be patients who will act like Dorothy and take his words a bit too literally.

It was, therefore, concerning to me that he took on the pharmaceutical industry from a position that can be viewed as inflammatory. Here were his "Four Secrets That Drug Companies Don't Want You to Know":

1. Drug companies underestimate dangerous side effects.
2. Drug companies control much of the information your doctor gets.
3. You're often prescribed drugs that you don't need.
4. Drugs target the symptoms, not the cause.

It was amazing to watch the reaction of the audience as these topics were discussed. Right from the outset of each discussion, the audience was clearly in agreement with Dr. Oz as evidenced by the nodding of heads and applause for some of his comments. To be clear, there were points that Dr. Oz made that I openly agreed with. In fact, each one of his "secrets" was rooted in a valid starting point. But overall, his views paralleled many preconceived incorrect notions, which continue to haunt the pharmaceutical industry. These need to be corrected before the pharmaceutical industry can regain trust as an innovator against disease.

DRUG COMPANIES UNDERESTIMATE DANGEROUS SIDE EFFECTS

Dr. Oz believes that all drugs have dangerous side effects. He is absolutely correct. It is impossible to discover and develop a drug that will be universally safe in millions of people around the world: males and females; old and young; large and small;

all members of different ethnic groups. Think about people with peanut allergies who can die by ingesting a single nut, or consider others who are lactose-intolerant. If basic foods are not tolerated by all, how can it be expected that a medicine, which is specifically designed to interact with critical biological processes in one's body, be totally safe for every person?

When a company wants a new medicine approved by the FDA, it files a New Drug Application (NDA). The NDA contains all relevant data from tests done with the potential new drug, starting with the earliest animal studies through all the human studies done to prove the efficacy and safety of the medicine. Thus, the FDA has access to all data generated with the drug, and they will determine, in conjunction with outside experts, the benefits of the drug along with its risks. If the drug is intended for use in patients for whom good treatment options already exist, the risks need to be relatively minimal because there is no need to expose patients to a new drug with added risks over existing, well-understood therapy. However, if the new drug is for a condition, especially a life-threatening condition, for which no good treatment option exists, the FDA will be tolerant of a certain degree of side effects in order to get such an important drug to patients in need. Thus, the risk–benefit profile of a drug is not evaluated via a fixed equation. It is based on evaluating the need of the new medicine as measured against the negative effects it might have for the patient.

So, if the FDA has such a rigorous system, how is it that drugs get approved that are later found to have more serious side effects than were originally known? In spite of the years devoted to studying a new drug in patients (often in excess of a decade) and the enormous costs involved (estimates will vary based on the nature of the drug, but in general anywhere from $1 to $3 billion is spent on a single new medicine), usually each NDA has test data from 2000 to 20,000 patients. Once the NDA is approved, it is likely that millions of patients will be prescribed the medicine. Thus, if the drug causes a rare side effect, say 1 in 50,000 patients, it may not be discovered until it has been broadly used. However, neither the drug maker nor the FDA is ignorant of such an occurrence. Every company has a pharmacovigilance program, the purpose of which is to monitor any unusual safety issues that may arise with both experimental and approved drugs. The FDA also maintains an Adverse Event Reporting System, a computerized database designed to monitor the safety of all medicines sold in the United States. The database contains reports submitted by manufacturers, health-care professionals, payers, and consumers, and these reports get scrutinized by FDA officials to determine if further investigation is warranted.

Despite this system, Dr. Oz is right in saying that we don't know the full profile of a new drug. However, his solution is to use more generic drugs:

> Use generic drugs, because they've been around for a long time, they're old drugs, and they've been used by so many people that we know about their side effects. And, they're cheaper.

Ironically, Dr. Oz opened his show talking about recent safety concerns that have arisen over the use of Alli, a generic drug. Clearly, Dr. Oz's suggestion that generic drugs are safer is an oversimplification. His contradiction highlights the fact that the public needs to be better informed about what a generic drug is.

When a drug company invents a new medicine and has it approved by the FDA, it has a patent on the new medicine that allows it a period of time (generally 10–12 years) during which it is the only company that can sell the drug. Once the patent for this medicine expires, it "goes generic"—that is, anyone can seek approval from the FDA to make and sell this drug—provided that their version of the drug is bioequivalent to the original medicine (that is, the drug behaves identically). In fact, the vast majority of prescriptions written every year are for generic drugs.

But just because a drug reaches generic status doesn't automatically make it universally safe. It just means that its patent has expired and it can be made by a lot of other companies. Because it has been on the market for a number of years, more is understood about the risk–benefit profile of the medication at this point than when it was first marketed. However, the safety profile of the compound doesn't change. As a generic, it still has the same side effects it had when it was a branded drug. A great example of this is acetaminophen, the active ingredient in Tylenol. There is no difference between Tylenol, the branded drug, and acetaminophen branded by CVS, Rite Aid, or another pharmacy. Acetaminophen has been taken by hundreds of millions of people around the world over the last 50 years. You can buy it in any drug store and most supermarkets. More is known about it than about most prescription medications. And yet acetaminophen is the leading cause of calls to Poison Control Centers every year (>100,000). Acetaminophen accounts for more than 56,000 emergency room visits, 2600 hospitalizations, and an estimated 458 deaths annually as a result of acute liver failure. These incidents are largely due to overdoses of this drug, both intentional and unintentional.

Perhaps the best explanation of the dangers of an old drug comes from David Stipp's article in *Fortune* magazine entitled: "Take Two Possibly Lethal Pills and Call Me in the Morning."[1]

In it, Stipp describes a medicine that is ". . . commonly given to patients for non-fatal conditions such as mild inflammation. Yet, studies suggest that it and several drugs like it are fatal to at least 10,000 Americans a year. The victims die grisly deaths, typically from internal bleeding." This drug is acetylsalicylic acid, also known as aspirin. This drug has been available over a century and has probably been ingested by over a billion people. And yet, if it were discovered in a pharmaceutical R&D laboratory today, it might never have been developed. The reason is simple. In standard animal testing, gastric lesions and bleeding are observed very quickly. Such toxicity might not be accepted by companies or regulators today because it would lead to thousands of patients being hospitalized from gastrointestinal bleeding. Thus, scientists would likely abandon aspirin and look for a safer compound that provided pain relief without this toxicity.

So, the bottom line is that every medicine can cause a safety problem in people. People should only turn to medicines as a last resort. Lifestyle changes, diet, and exercise are the sorts of things people should do before turning to pills. And when people do need to turn to medicines to alleviate their illness, they should discuss with their doctor the potential side effects that the prescribed medication has. Yet, despite all of these caveats, medicines save lives, prevent serious debilitating diseases, and improve the lives of millions of people each year.

After reading all of this, a person may be turned off from taking any medicine, be it prescription or generic. Why ingest anything that might cause harm? However, people do get sick, get headaches, sprain ankles, and so on. Interestingly, a number of people turn to herbal remedies for these situations.

There is no denying that many drugs have their beginnings in traditional medicines. Tales abound of the healing properties of Chinese herbal medicines or plant treatments used by Native American healers. Hundreds of years ago, people knew that chewing the bark of a willow tree relieved their pain. In the 1800s, salicylic acid was identified as the active component in willow bark; and, in 1899, scientists at Bayer synthesized a new form of salicylic acid called acetylsalicylic acid, the aforementioned aspirin. This is not an isolated example.

The modern drug industry may owe its roots to early traditional medicine, and certainly the end goal is the same now as it was hundreds of years ago. But the continued grip of herbal medicines on the general population is astounding. The *New York Times*[2] reported that in 2011 the US sales of dietary supplements purported to be used for health benefits came to $28.1 billion, a jump from $21.3 billion in 2005. Consumers of these products are driven to them because of the expectation that natural products are inherently safer than medicines derived from drug companies. Furthermore, there is a belief that centuries of use ensure efficacy, a belief echoed somewhat in Dr. Oz's views of generic medicines. However, studies show that this is not always true.

Ginkgo biloba is marketed as an herbal medicine that enhances memory, and it may be the most widely consumed herbal treatment used to prevent age-related cognitive decline. Its popularity is attested to by its US sales, which exceed $250 million annually. In 2008, the *Journal of the American Medical Association* (*JAMA*) published the initial results of the Gingko Evaluation of Memory (GEM) trial,[3] a randomized, double-blind, placebo-controlled clinical trial designed to test *Ginkgo biloba* for preventing dementia. GEM was conducted in five academic medical centers in the United States between 2000 and 2008 with 3069 volunteers aged 75 or older with normal cognition or mild cognitive impairment. Half of this population received a standardized extract of *Ginkgo biloba* and half received a placebo; they were followed for an average of 6.1 years. At the end of this time, there was no statistically significant difference in the occurrence of either dementia or Alzheimer's disease between the two groups. In other words, *Ginkgo biloba* had no beneficial effect with respect to these parameters.

Saw palmetto is extracted from the fruit of *Serenoa repens*. Native Americans use the saw palmetto berries to treat urinary problems, and it has grown popular among older men as a treatment for benign prostatic hyperplasia (BPH). In 2006, the results of the Saw Palmetto Treatment for Enlarged Prostates (STEP) trial were published.[4] This was another multicenter double-blind, randomized, placebo-controlled trial comparing 160 mg twice a day of saw palmetto versus a placebo. Despite looking at a variety of parameters (prostate size, residual volume after voiding, PSA score, or quality of life), after one year there was no difference between the results seen with saw palmetto and placebo. In addition, the results of yet another trial, this one published in *JAMA*,[5] showed that doubling and then tripling the dose used in the STEP trial for two years also had no effect. Saw palmetto is no more effective than placebo in treating BPH.

The following headlines are pretty impressive:

"Smokers quit with cheap Bulgarian remedy. For as little as $6, there may be a smoking-cessation remedy that actually works."

"Soviet-era pill from Bulgaria helps smokers quit."

"Smokers get chance to beat the habit with 12-pence tablets."

This publicity was generated by a paper in the *New England Journal of Medicine* (*NEJM*) entitled "Placebo-Controlled Trial of Cytisine for Smoking Cessation."[6] In this double-blind study, 740 smokers were equally divided into two groups: one taking a cytisine preparation sold in Bulgaria as Tabex, the other given placebo. After 4 weeks of dosing, treatment was stopped and the patients were monitored for a year, after which time 8.4% (31 participants) on cytisine and 2.4% (9 participants) on placebo had quit smoking. How does this compare with other smoking cessation regimens? A similar type of smoking cessation study published in *JAMA*[7] showed that Chantix (varenicline) provided 23% continuous abstinence, buproprion 14.6%, and placebo 10.3%. (In terms of full disclosure, Chantix was discovered and developed at Pfizer during my tenure as head of research.)

The cytisine study is noteworthy in that, while its anti-smoking effects have been known for over 40 years, this is the first reported clinical trial done in a double-blind placebo-controlled manner as required by regulatory agencies like the FDA. But the real driver and interest in this work is the fact that the Tabex brand of cytisine is cheap and available online. The leader of the *NEJM* study, Professor Robert West, predicted that "the publicity surrounding his findings would trigger a surge in people turning to websites to obtain it." Costs of smoking cessation products vary from country to country; but, as the study authors point out, the cost of a course of cytisine therapy is about 10–20% that of Chantix. That difference could mean big savings to agencies like Britain's National Health Service and Medicare in the United States.

So should we rush to get this drug? I would argue not just yet. There are still a few issues that need to be addressed. First of all, how effective is cytisine compared to the currently marketed agents? Unfortunately, the *NEJM* study does not include a "positive comparator," either Chantix or buproprion, so that one can get a sense of how the efficacy of a current treatment compares directly with cytisine in the same study. Looking at the data, it doesn't seem that cytisine is as effective as the other compounds. If cytisine is only half as effective in causing people to quit smoking, that should cause payers, physicians, and patients to think twice about replacing other methods with cytisine.

However, there is a much bigger issue facing cytisine use: safety. While Tabex has been available in all former socialist countries since the early 1960s, it was withdrawn from the market in many of these countries when they joined the European Union. Safety concerns could possibly have been one reason. The Tabex preparation originates from the plant *Cytisus laborinum* L. There are reports of people getting poisoned with the seeds of this plant. J. F. Etter, in the review article entitled

"Cytisine for Smoking Cessation,"[8] states that "Poisoning in children who eat *laborinum* seeds is frequent" and that "in an average summer over 3000 children are admitted to hospitals in England and Wales because of *laborinum* poisoning." The symptoms, which include nausea, abdominal pain, respiratory stimulation, and muscle weakness, are consistent with poisoning symptoms with nicotine, which is related chemically to cytisine. Clearly, eating these seeds results in a large overdose of cytisine, and the Tabex dose levels of cytisine are much lower. But this points out the need for extensive safety studies of cytisine. The side effects of Chantix and bupropion are well known. But these adverse effects were found as a result of the extensive data safety monitoring that has been accumulated by the manufacturers over the years these drugs have been on the market. To my knowledge, such an adverse event monitoring system has not been in place for Tabex.

Peter Hajek, director of the Tobacco Dependence Unit at Queen Mary University Hospital in London, said the following to the Associated Press:

> It is possible that extensive bureaucracy and over cautious regulations will prevent its (cytisine's) use in the U.S. and Europe.

One would hope that a person in his position would be concerned with the safety and efficacy of a drug before advocating its use.

There is no doubt that cytisine, in the form of Tabex, is a cheap way to help some people stop smoking. But is it more effective and safer than existing medicines? That can't be answered with the current data; more studies are needed. Furthermore, by the time these types of clinical trials are carried out, it is possible that a generic form of Chantix, varenicline, will be available, making the cost arguments moot.

One may ask: What's the harm in these herbal remedies? These extracts aren't overtly toxic, and so people should be free to take whatever they'd like. Maybe taking these herbal medicines gives them peace of mind. However, because herbal drugs aren't nearly as well studied as prescription drugs, there could be issues with them that are unrecognized. For example, does taking either *Ginkgo biloba*, saw palmetto, or cytisine interfere with the metabolism of other medications that one is taking? This idea isn't far-fetched. It is known, for example, that St. John's wort, a controversial herbal medicine taken by people to treat their depression, limits the effectiveness of a variety of medications, including antiviral agents, birth control pills, and some anticancer medications.

If a person wants to take herbal medications, that's his or her personal choice. But patients should be aware that these medications haven't been subjected to the rigorous regulatory agency scrutiny that all prescription medicines must receive. Buyer beware.

DRUG COMPANIES CONTROL MUCH OF THE INFORMATION YOUR DOCTOR GETS

Another topic closely related to the safety of drugs is the claim that drug companies hide negative data on their experimental medicines. These could be data that show that the drug works poorly or that the drug has unreported side effects. This is a tenet of Dr. Abramson, who made the following comment during our discussion:

> About 85% of clinical trials are now funded by the drug industry. They own that data. The docs don't understand that they are getting a selected, filtered version of what the information is.

Some of Dr. Abramson's charges have a factual basis. In fact, most clinical trials *are* funded by the drug companies. They are the ones studying the investigational new drug, and so they are obliged to pay for the costs of running these trials. The trials themselves, however, are run by independent physicians at various academic centers and teaching hospitals around the world. Thus, it is safe to assume that they are rigorous in the conduct of the studies as well as in reporting all of the beneficial and harmful effects that the new medicine may cause. As was stated earlier, these data get reported to the supervising regulatory agency.

Furthermore, it is no longer possible to hide clinical trials from the public. As part of the established principles for good conduct of clinical trials, a summary of the protocol of each trial must be made available both while the study is being recruited and while the study is ongoing. The study results then must be made available to the public in a timely fashion. All of this occurs by making use of a government hosted website: www.ClinicalTrials.gov. All companies that run a trial, be it with a drug or a device, must register the trial on this website, making the clinical activities of a company very transparent.

So why is Dr. Abramson still concerned about this? He would like all clinical trial results published on ClinicalTrials.gov and in peer-reviewed journals. This, he believes, would make these data more readily available without the need to go searching for them. The problem is that companies don't publish all of their results. This is not because they are trying to hide data. Rather, not all trials are worthy of scientific publication, and the decision whether or not to publish a manuscript rests with the medical journal's editorial staff. A great example exists in the field of depression—interestingly, a topic also discussed by Drs. Oz, Abramson and myself on the show.

Major depression is a serious disease. In 2005, the *New England Journal of Medicine* published Dr. J. John Mann's excellent review entitled "The Medical Management of Depression."[9] In this piece, Dr. Mann states: "Major depressive disorder accounts for 4.4% of the global disease burden, a contribution similar to that of ischaemic heart disease. . . ." Equally concerning is that: "Patients who have diabetes, epilepsy, or ischaemic heart disease with concomitant major depression have poorer outcomes than do those without depression."

One would think, therefore, that those concerned with medical care would be grateful for the availability of drugs to treat depression. Yet, there are many who challenge the value and the need for antidepressants. Dr. Abramson, in his *Overdosed America*, says the following in his book: ". . . new antidepressants were found to be not even 10% more effective than placebos: symptoms of depression improved by 30.9% in the people who took placebos and by 40.7% of people who took the newer antidepressants." Dr. Abramson's implication is that these drugs offer little value over sugar pills.

Dr. Marcia Angell, a noted pharmaceutical industry critic and a former editor of the *New England Journal of Medicine*, has also expressed a number of concerns about antidepressants, not the least of which is that drug companies rarely publish negative data on these drugs. In her article entitled "The Epidemic of Mental Illness—Why?," which appeared on June 23, 2011 in the *New York Review of Books*, Dr. Angell states the following:

> When drug companies seek approval from the FDA to market a new drug, they must submit to the agency all clinical trials they have sponsored. . . . If two trials show that the drug is more effective than a placebo, the drug is generally approved. But companies may sponsor as many trials as they like, most of which could be negative—that is, fail to show effectiveness. . . . For obvious reasons, drug companies make very sure that their positive studies are published in medical journals and doctors know about them, while the negative ones often languish unseen within the FDA. . . .

Factually speaking, Dr. Abramson and Dr. Angell are correct. Antidepressant drug trials usually show only modest superiority of the experimental drug over the placebo.

And yes, drug trials where there is no difference in the efficacy of the drug versus the placebo normally aren't published. However, such statements don't tell the whole story in the battle to treat psychiatric disorders.

In doing clinical studies, there are some medical areas where it is easy to measure whether a drug is working or not. If you have a new antibacterial to treat an infection, you can take blood samples to measure the effect that the drug is having on the eradication of the bacteria from the patient. Similarly, with a drug to treat high blood pressure, you can treat a patient with a new drug and take real-time blood pressure measurements to quantify the drug's effects.

Psychiatric diseases are different. In clinical studies, patients are given a standardized test, such as the Hamilton Rating Scale for Depression (HRSD), which involves answering a variety of questions about a patient's mood, anxiety, ability to sleep, and so on. The patients are then randomized to receive either the placebo or the test drug. Each week, the patient returns and is seen by the psychiatrist to determine if any improvements are evident. Generally, these studies last for about two months. Such studies are notorious for the high efficacy response rates seen in patients who turn out to be taking the placebo. In an excellent 2002 paper on this subject, "Placebo Response in Studies of Major Depression,"[10] analysis of 75 clinical trials showed that the response to the placebo across the trials ranged from 10% to more than 50%. While the authors couldn't point to one definitive reason why such high placebo rates occur, they offered a few possibilities:

1. Patients in these trials, whether on the placebo or on the drug, are usually allowed to take sedatives and anti-anxiety medication, so the placebo responses encompass the effects of these drugs.
2. The weekly physician visits contribute to the patient's great sense of improvement.

3. It's likely that these studies included patients with milder, briefer, and more responsive forms of depression, thereby enhancing the chances of either the placebo or the drug being effective.

Thus, it is well established that there are high placebo efficacy rates in clinical trials in psychiatric disorders. As a result, studies where the efficacy of the placebo is equal to that of the drug being tested are rarely accepted for publication by reputable scientific journals. It is not that companies are hiding data. Rather, such a result is of little interest. Most skilled in the science behind the results realize that seeing the benefit of a placebo over an experimental drug is a hazard of this type of study. However, it is big news when a new antidepressant does, in fact, show statistically significant efficacy over the placebo. This instance is deemed very important by the scientific community; and, therefore, medical journals are very willing to publish such results. In other words, Dr. Angell's concerns are unjustified.

The placebo effect is not unique to depression. Shirley Wang, in an excellent *Wall Street Journal* article entitled "Why Placebos Work Wonders,"[11] gives the following examples, one in weight loss and another in fertility.

> Hotel-room attendants who were told they were getting a good workout at their jobs showed a significant decrease in weight, blood pressure, and body fat after four weeks, in a study published in *Psychology Science* in 2007 and conducted by Alia Crum, a Yale graduate student, and Ellen Langer, a professor in the psychology department at Harvard. Employees who did the same work but weren't told about exercise showed no change in weight. Neither group reported changes in physical activity or diet.

> Fertility rates have been found to improve in women getting a placebo, perhaps because they experience a decrease in stress. A recent randomized trial of women with polycystic ovarian syndrome found that 15%, or 5 of 33, got pregnant while taking placebo over a six-month period, compared with 22%, or 7 of 32, who got the drug—a statistically insignificant difference. Other studies have demonstrated pregnancy rates as high as 40% in placebo groups.

Given this background, it becomes clear that, while the results of all clinical trials need to be registered and filed with regulatory agencies, it is hard to justify publishing all of them. This is not to say that publishing negative data for a medicine isn't of value. Studies that show a drug doesn't work in certain indications are just as important as those that show a benefit. But the fact is that the results of a lot of studies don't merit publishing in major medical journals, be it due to placebo effects or other reasons. This isn't burying negative data. It is simply publishing meaningful data, be it positive or negative, that can help guide physicians in properly treating their patients.

However, Drs. Abramson and Angell have a valid criticism in terms of the timing of publishing the study results. It is great to have every clinical trial registered on ClinicalTrials.gov. at the time the first patient is enrolled in a study. However, once the study is completed and fully analyzed, it is crucial for the results to be reported in a timely fashion. There are two articles in the *British Medical Journal* (*BMJ*) that suggest that this is not happening. In 2007, the FDA began to require that summary results for all trials be posted on ClinicalTrials.gov within 12 months.

This was made mandatory for trials started or ongoing as of September 2007. So, how are companies doing?

In the first *BMJ* paper,[12] the authors looked at studies that were already registered on ClinicalTrials.gov and completed between January 1 and December 31, 2009. They found that only 22% had reported results within one year of completion (another 10% were not subject to mandatory reporting). The second *BMJ* paper[13] looked at NIH funded (academic-based research) trials. Interestingly, these authors found that for academic trials carried out from September 30, 2005 until December 31, 2008, only 46% had published results 30 months after the completion of the trial. Clearly, both the industry and academic groups are not distinguishing themselves in the timely publication of trial results. The fact that both research segments are having difficulties complying with this FDA mandate leads one to speculate that there may be unforeseen inherent issues with the specific time requirements.

Nevertheless, this is clearly a black eye for an industry trying to regain the public's trust. In an accompanying *BMJ* editorial, Richard Lehman and Elizabeth Loder make the following statement.

> What is clear from the linked studies is that past failures to ensure proper regulation and registration of clinical trials, and a current culture of haphazard publication and incomplete data disclosure, make the proper analysis of the harms and benefits of common interventions almost impossible for systematic reviewers. Our patients will have to live with the consequences of these failures for many years to come.

What is at stake here is the credibility of the industry. Pharmaceutical companies claim to be committed to transparency, and perhaps there is no greater place to improve the pharmaceutical company image than by the timely publication of clinical trials data. Not doing so gives critics a major target because it enables them to continue the charge that companies are hiding negative data on their new medicines. I don't believe that is the case. But this problem needs to be corrected if the industry is going to have any credibility on this topic.

YOU'RE OFTEN PRESCRIBED DRUGS THAT YOU DON'T NEED

A major criticism of drug companies is that they try to encourage people to take pills that often aren't needed or for indications that are nonimportant. As was stated earlier, high on the radar screen for Drs. Abramson and Oz are statin drugs—compounds like Lipitor (atorvastatin) and Zocor (simvastatin), which are designed to lower LDL cholesterol, the so-called bad cholesterol, which is associated with atherosclerosis leading to heart attacks and strokes. They feel that these are far too often prescribed to patients who are basically healthy.

Drs. Abramson and Oz are not unique in their view. Some physicians and cardiovascular experts believe that statins are lifesaving drugs. Yet, there are those who believe they should only be used in cases when heart disease is clearly evident. The evidence, however, clearly demonstrates the value of statins and clearly measures their risk–benefit profile.

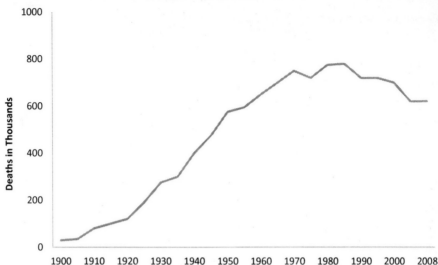

Figure 1.1   Deaths attributable to diseases of the heart in the United States (1900–2008). From the American Heart Association 2012 Statistics Update.

First of all, while you will see that I am a big proponent of statins, I am also a big believer in diet and exercise as a first step in everyone's personal health regimen. This is not true just in treating high cholesterol. The benefits of a healthy diet and an active lifestyle extend to diabetes, osteoarthritis, osteoporosis, and even psychological disorders like depression. But oftentimes, diet and exercise are not sufficient to reduce the risk of these diseases; and, at some point, specific medicines may be required to restore a person's health or to prevent long-term consequences of the disease. Furthermore, I practice what I preach. I work out for an hour every day. I do this for a variety of reasons, not the least of which is that I have a strong family history of heart disease. I believe that my exercise regimen will minimize my cardiovascular risk.

The American Heart Association has been monitoring deaths due to cardio-vascular disease (CVD) in the United States for over a century. While the CVD death rate grew steadily for most of the twentieth century, it leveled off and then began to drop somewhat over the past 25 years (Figure 1.1). Nevertheless, CVD is still the leading cause of death in the United States, with 600,000 people dying annually, which accounts for more than 25% of all deaths in the country. The direct costs associated with treating heart disease amount to over $80 billion/year, and indirect costs attributed to loss of productivity exceed $60 billion/year.

Despite the progress made in moderating the CVD death rate, it is still a major disease. Furthermore, as the obesity epidemic continues in the U.S., recent headway made in the CVD death rate is liable to be counteracted by the increase in obesity, which is already resulting into a concomitant increase in type 2 diabetes, a precursor to heart disease (Figure 1.2). Even with improvements in diagnosis and treatment, better understanding of risk factors, reductions in smoking, etc., CVD is going to remain a major health problem for decades.

Age-adjusted Percentage of U.S. Adults Who Were Obese or Who Had Diagnosed Diabetes

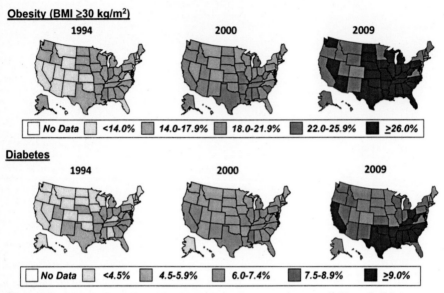

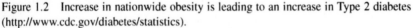

Figure 1.2 Increase in nationwide obesity is leading to an increase in Type 2 diabetes (http://www.cdc.gov/diabetes/statistics).

Yet, diet and exercise are not always a panacea. Despite the obvious benefits, two studies in the *New England Journal of Medicine* suggest that for the majority of the obese population, just diet and exercise won't be enough.

The papers are entitled "Comparative Effectiveness of Weight-Loss Interventions in Clinical Practice" and "A Two-Year Randomized Trial of Obesity Treatment in Primary Care Practice."[14,15] Essentially, these papers are similar in that they look at the effectiveness of different interventions in primary care practices by physicians who were trying to help their obese patients gain better control of their health. On average, the patients in these studies had a body mass index (BMI) of about 35 (e.g., a height of 5'7" and a weight of 220 pounds) and had at least one cardiovascular risk factor (high blood pressure, plasma glucose, or cholesterol). Both studies had control groups who received usual physician care. In the behavioral intervention study, besides the control group, one group received additional face-to-face counseling and the other group received advice remotely (telephone or email). In the obesity treatment study, the control group was compared to two groups: those who received monthly lifestyle counseling and those who received enhanced counseling plus meal replacements and weight-loss medications.

The good news is that, in both studies, those patients who were getting enhanced treatment, be it extra counseling on behaviors, more frequent sessions with their doctors or physician assistants, or enhanced lifestyle counseling, all had sustained statistically significant weight loss after two years. The amount of weight loss wasn't trivial—around 5%. Even if the counseling was done remotely, the results were meaningful. Thus, extra time spent by primary care physicians and their associates can make a difference in helping their patients lose weight.

But the disappointing news is that even with the loss of 5% of body weight, these patients are still obese. Someone who loses 5% from his or her 220-pound frame now weighs 209 pounds with a BMI of 34—still well in the obese range. When you consider that the Center for Disease Control statistics for 2010 show that there are now 12 states where more than 30% of the adult population is obese, the loss of 5% body weight is just a small step to where we must go in order to improve the nation's health.

The impact of obesity on the future health of the United States cannot be trivialized. Lifestyle changes are very important and can't be minimized. But for millions of Americans, this isn't nearly enough.

One might argue that these people may be obese; but, if they are feeling fine, why should they be put on medications? Chances are that even if they haven't progressed to diabetes, these people have high blood pressure and high cholesterol; and, according to work published in the *New England Journal of Medicine*,[16] if at age 45 you have two or more of either elevated blood pressure, cholesterol, diabetes or smoking, and if you are a man, you have a 50–50 chance that you will have a stroke or a heart attack at some point in your remaining lifespan. Even a 45-year-old woman with two of these risk factors has a 30% chance of such a cardiovascular event in her lifetime.

So, let's say these data convince you to watch what you eat, go for vigorous walks a few times a week, and stop smoking. You're feeling better and you are no longer obese but you are still overweight and your cholesterol is still high. Some physicians will prescribe a low-dose statin for you, particularly if you have a family history of heart disease. But not all will. Why?

Physicians reluctant to prescribe statins tend to have two fundamental questions: Are their protective benefits maintained over many years of use? Are patients exposed to these drugs for long periods susceptible to other diseases? Over the past few years, studies have shown that statins like Lipitor (atorvastatin) can, in fact, reduce the occurrence of heart attacks and strokes in patients with heart disease. A study reported in *The Lancet* from the Heart Protection Study (HPS) Collaborative Group provides answers to these questions.

The trial design was pretty simple. This group of patients at high risk of heart attacks and/or strokes was given either 40 mg of simvastatin or placebo for slightly over 5 years, and then post-trial follow-up occurred for another 5+ years, bringing the duration of the study to 11 years. Not surprisingly, those patients on simvastatin had an average decrease in LDL cholesterol of 40 mg/dL and, more importantly, a decrease of 23% in major vascular events. This benefit continued throughout the post-trial follow-up period. Equally important was the fact that there was no evidence of "emerging hazards" (e.g., cancer) resulting from long-term simvastatin use.[17]

The lead author of the HPS, Dr. Richard Bulbulia of the University of Oxford, commented: "All of those at increased vascular risk should start taking statins early and continue taking them long-term." He also commented that these results should provide reassurance to patients and physicians about the safety of statins and that the results should translate to other members of this class of medicines such as atorvastatin and Crestor.

The HPS has provided valuable results at a time when health-care providers are struggling with increasing rates of obesity, diabetes, and heart disease. But this study is important for another reason. Medications designed for controlling a chronic disease that patients will need to take for decades—not only heart disease, but also diseases like osteoporosis, depression, or even cancer—will need this type of long-term outcome study, not just to provide patients and physicians with all-important risk–benefit data, but also to justify to payers the value of new medications. These studies add time to the development of a new drug and greatly increase development costs. But they are invaluable for teaching physicians and patients the long-term implications of taking a drug like a statin.

Despite these results from the Heart Protection Study group, patients may seek a more "natural" medication to treat their cholesterol abnormalities. A way one can do this is by using niacin. Niacin, also known as vitamin B3, is known to raise HDL, the so-called good cholesterol, by about 25% as well as modestly lower both LDL and triglycerides. It has been used for decades to treat dyslipidemia based on results from the Coronary Drug Project (CDP). Carried out in the late 1960s, the CDP study tested niacin versus placebo in men who had a previous heart attack, over a period of five years. Interestingly, niacin showed no difference from placebo in the death rate of the men in this study, but fewer patients on niacin had a nonfatal heart attack or stroke, by 26% and 24% respectively. This study is the basis of the use of niacin in cardiovascular (CV) disease.

In fact, I tried niacin as a way to control my cholesterol levels. Despite having a normal BMI, normal blood pressure, and daily exercise, I have high LDL cholesterol. Given my family history of heart disease, I decided to try niacin to get my LDL levels to the recommended American Heart Association levels.

My experience with niacin was pretty typical. There were modest reductions of both total cholesterol and LDL (~15%), but these changes weren't maintained over time. But I also experienced the major niacin side effect, flushing. This irritation wasn't minor. The flushing was intense and was accompanied by itching and heat sensations. Because of this side effect, many patients refuse to stay on this medication despite its potential benefits. After about a year, my physician took me off niacin and I started on Lipitor, which was far more effective for me than niacin and which I tolerated very well.

So, why am I giving you this personal history? A study reported in the *New England Journal of Medicine*,[18] along with an accompanying editorial, call into question the value of using niacin to treat CV disease. The AIM-HIGH trial, co-sponsored by the NIH and Abbott, looked at patients with established CV disease who were already on intensive statin therapy. The goal of this study was to see whether adding niacin therapy provided any extra benefit. The rationale for this was pretty sound. Unlike statins, niacin can significantly raise HDL and further lower

LDL. Shouldn't combining both modalities work better? Surprisingly, it didn't. While the expected beneficial changes in terms of raising HDL did occur, adding niacin to intensive statin therapy was no different from adding a placebo in terms of preventing heart attacks, strokes, or other adverse CV events.

The *NEJM* editorial accompanying the AIM-HIGH study results entitled "Niacin at 56 Years of Age—Time for an Early Retirement?" basically questions further use of niacin given the copious data with statins showing the superiority of these drugs in CV disease therapy. This is causing some intense debate amongst cardiologists, who are unwilling to give up on niacin after this one study. The defenders of niacin correctly point out that there are other long-term studies with niacin currently underway that will provide a more definitive answer to the value of niacin for treating heart disease.

Niacin is a medicine that has been used by physicians for 56 years. Physicians take comfort in the fact that it has been around for so long and it has been taken by millions of people, so they know what the side effects are. Yet, niacin hasn't been as intensively studied as newer classes of lipid modulating drugs. It is now being subjected to the same type of scrutiny demanded by the FDA of new drugs.

As was said earlier, decades of use does not ensure that a medicine is automatically safe and/or effective. Industry detractors seem to forget that pharmaceutical companies are full of people who also need medicine. I was my own case study in the effectiveness and risk–benefit profile of Lipitor versus niacin. For me, Lipitor was the answer. Whether or not that is also the case for others is a decision that a patient must make in consultation with his or her physician. However, one thing is for certain: Only long-term, well-controlled studies can provide assurance that a medicine is both safe and effective.

So, if your doctor recommends that you begin taking a statin, is he or she prescribing a drug for you that you don't need? Based on recent clinical studies, this is not the case if diet and exercise have not completely solved your health concerns and you still have significant disease risk factors. However, pharmaceutical industry critics still believe that this type of medicine is "disease mongering." This will be addressed next.

DRUGS TARGET THE SYMPTOMS, NOT THE CAUSE

This was the fourth "secret that drug companies don't want you to know." First of all, it is untrue. Anyone who has needed an antibiotic knows that these medicines treat the cause of the illness by killing the bacteria that cause one's fever, aches, and so on. The same can be said for people with AIDS, who take medicines discovered and developed by the pharmaceutical industry and which keep their disease under control. The same can be said for drugs that treat cancer or that enable one to quit smoking. Even Viagra can be said to treat a cause and not a symptom.

The fourth "secret," however, has a different purpose than just attacking drugs that treat symptoms. After all, people take drugs to deal with symptoms routinely, such as ibuprofen for a headache or antacids for heartburn. But the intent of people who allege this is based on the premise that drug companies hype up conditions that really aren't diseases in order to make money. Worse, they charge that the industry invents diseases.

The view that scientists in a pharmaceutical company sit around dreaming up new diseases and then convince people that their minor ailment urgently needs drug treatment is absurd. First of all, a company cannot simply declare a new disease and market a drug to treat it. A disease must be recognized by global regulatory agencies which set up criteria that a drug must meet in order to have even the most remote chance to be approved. Second, payers must believe that the condition is serious enough to warrant reimbursement of the cost of the drug to treat it. Third, physicians must believe the disease is serious enough to be willing to prescribe a drug to their patient to treat it. And finally, patients must be concerned enough about their pain or discomfort to be willing to seek treatment in the first place. Thus, in order for the "world disease mongering conspiracy" to be successful, patients, physicians, payers, and regulators must act in concert with the exploitative drug companies. Doesn't this seem the least bit far-fetched?

Yet, what most people might consider preventative medicine, others regard as "disease mongering," as exemplified by the following quote from the *British Medical Journal.*[19]

> Some forms of "medicalisation" may now be better described as "disease monger-ing"—extending the boundaries of treatable illness to expand markets for new products.
>
> —Ray Moynihan

Moynihan is a layman and one of the leading critics of the pharmaceutical industry. He has a strong belief that the industry effectively invents diseases by the medicalization of conditions in such a way that convinces healthy people that they are sick. One example used in this paper is in the area of osteoporosis:

> Like high blood pressure or raised cholesterol levels, the medicalisation of reduced bone mass—which occurs as people age—is an example of a risk factor being concep-tualized as a disease. . . . Slowing bone loss can reduce the risk of future fracture—just as lowering blood pressure can reduce a person's chance of a future stroke or heart attack—but for most healthy people, the risks of serious fractures are low and/or distant, and in absolute terms, long-term preventive drug treatment offers small reduc-tions in risk.

Is bone-thinning a disease? Of course not. But, if you are a petite female of Asian or northern European background, bone thinning is the first sign of osteopo-rosis. Unfortunately, when journalists like Moynihan make statements like this, it results in people ignoring health issues and not taking the steps necessary to forestall diseases like osteoporosis. Thus, it was refreshing to see Jane Brody's view[20] in a *New York Times* article entitled "A Reminder on Maintaining Bone Health."

Brody believes that osteoporosis is underdiagnosed because of a reluctance to get bone density tests and is undertreated because people avoid drug therapy for fear of side effects. At age 60 she was found to have osteopenia, a condition character-ized by low bone density but without frank osteoporosis. She likens osteopenia to prediabetes or prehypertension. At this stage, one doesn't need to take drugs but lifestyle changes are recommended, such as regular weight-bearing and strength-training exercise, intake of calcium and vitamin D, smoking cessation, and limited alcohol consumption. However, people with osteopenia can benefit from drug therapy if they have already had a fracture.

But what about the safety of drugs for osteoporosis? Brody does an excellent job in discussing the risk–benefit profile of bisphosphonates, the major class of drugs prescribed for this disease.

> On average, the bisphosphonates reduce the risk of a fragility fracture by 30–50%. By comparison, the risk of the most talked about serious side effect—an atypical fracture of the femur, or thigh bone—is minuscule.

What appealed to me about this article is the fact that an independent journalist from the *New York Times* has basically pointed out that bone loss is a real issue and needs to be treated long before symptoms arise. Brody's messages are pretty clear: Monitor your bone health; at the first serious signs of bone loss, you should make lifestyle changes; if your bone loss evolves into early osteoporosis, you should work with your doctor to identify the medicine best suited for you. Unlike Moynihan's minimization of a common disease of aging women and even some men, Brody has provided a thoughtful commentary on how best to approach bone loss. It is great advice.

Another term used by industry critics is "disease creep," which was used in an essay by Jeanne Lenzer entitled "Disease Creep: How We're Fooled into Using More Medicine than We Need."[21] Lenzer's views can be summarized in her quote below:

> Elevated cholesterol is not a disease. It doesn't cause symptoms. It is a risk factor. People with high cholesterol levels are somewhat more likely to develop a heart attack or stroke, but they are at far less risk than individuals who already have cardiovascular disease. This is the definition of disease creep: when pre-conditions or risk-factors are treated as if they are the same as the actual disease state.

In Lenzer's utopia, you wouldn't get a statin until *after* you have already had a heart attack. The problem is that many first heart attacks are fatal—you don't get a second chance to go on statin therapy then. She is correct in saying that just having high cholesterol alone does not justify taking a statin to prevent a heart attack or stroke. But CVD risk factors also include male sex, older age, family history of heart disease, post-menopause, smoking, obesity, high blood pressure, diabetes, and stress. If a patient presents to a physician with multiple risk factors and if diet and exercise have not been effective in lowering cholesterol levels to those recommended by the American Heart Association, that physician would be remiss if the patient wasn't prescribed a statin. Waiting for a patient to first have a heart attack or stroke before providing such treatment would be irresponsible.

Lenzen implies in her article that the prophylactic use of statins may only prevent 1 in 50 heart attacks. I don't necessarily agree with that number, but let's say that is correct. There are 785,000 first heart attacks per year in the United States. Even employing Lenzen's assumptions, the use of statins in the overall treatment paradigm of patients with multiple CVD risk factors would prevent thousands of heart attacks or strokes annually. Now that the most-studied statins like simvastatin and atorvastatin are generic, it would seem like the cost–benefit of statin use to prevent first heart attacks is noncontroversial. This isn't "disease creep"—it is simply good medical practice.

What came as a big surprise to me in discussing "The Four Secrets That Drug Companies Don't Want You to Know About" is the little respect that doctors are given in the health-care process.

I have a tremendous respect for doctors. It is incredibly difficult to get into medical school, and so the academic credentials of medical students are stellar. Then, they go through four years of medical school and another 3+ years of residency and then another year or more of study on a fellowship in their specialty. Only very talented and dedicated people get to be a part of this profession. And yet, to hear drug critics talk, you'd think that doctors have little knowledge of diseases and that they are at the mercy of a drug rep to teach them the latest medicine. The critics make it seem that doctors are only too eager to prescribe a new medicine especially when a drug rep offers them a free pen and pizza for lunch. Finally, another misconception is that doctors apparently turn weak-kneed when a patient comes in and asks for a new drug because they are unable to say no to such requests.

Well, my experience is a lot different. Doctors are very skeptical of new medicines. And why shouldn't they be? They are smart individuals who have a great deal of experience in treating their patients. They attend scientific meetings in their specialty and read the latest medical journals. They know the strengths and weaknesses of the medicines they prescribe, and they know their patients' conditions and needs. They also know that a major priority is "to do no harm." Why would they switch to something new when they are already successfully treating their patients with effective medicines?

Doctors seek new medicines for conditions where no treatment exists. They will be interested in a new medicine when current treatments do not work for some of their patients. They will also be interested in a new drug if it has advantages over older ones. But they are the gatekeeper for the medicines that their patients get. It is not the drug company, it is not the drug rep, it is not really the FDA. Ultimately, doctors decide in consultation with their patients. It is this relationship that determines if new medicines have value. No amount of advertising or free drug samples should change this. To imply otherwise is to insult the entire medical profession.

CONCLUSION

My experience with *The Dr. Oz* show was very worrisome. To hear Dr. Abramson criticize the biopharmaceutical industry with partial facts was concerning enough. However, watching the audience completely buy in to the "four secrets" was alarming. It was clear that these people had no clue about the value that the biopharmaceutical industry brings to the world's well-being.

President Obama signed the National Alzheimer's Project Act into law in 2011. He has publicly stated that our national goal should be to cure this disease by 2025. Where do people think the experimental drugs to treat Alzheimer's disease are coming from? Who does the public think will pay for the bulk of the clinical trials to prove or disprove whether the compounds designed to treat this disease actually work? Who will run the safety studies to show that such drugs will be safe for use in the general elderly population? Great research is done at the National Institutes of Health. Great fundamental research occurs daily at the universities and research institutes around the world. But the actual translation of this work into medicines occurs only in biopharmaceutical companies. The general public doesn't have any appreciation of this.

At times, the industry doesn't help itself. There is certainly a need for greater transparency in the R&D process, particularly with regard to clinical trial results. The industry hasn't done a good job in educating the public on the contributions it has made in the battles against heart disease, AIDS, cancer, and so on. In addition, I personally wonder about the wisdom of direct-to-consumer advertising.

But this industry needs to be supported and appreciated by the public. All of us benefit from the R&D output of a healthy and robust biopharmaceutical industry. Attempts to belittle it with people talking about "The Four Secrets That Drug Companies Don't Want You to Know" impact the long-term health of us all. Rather than fear these secrets, people should worry more about the industry's productivity. Beyond the need for drugs to treat Alzheimer's disease, patients are awaiting breakthrough medicines in diabetes, psychiatric disorders, cancer, antibiotic-resistant infections, and so on. For such breakthroughs to occur, biopharmaceutical R&D has to function optimally. Yet, industry productivity has sagged in the last decade. What is the cause of this drop? Can it be fixed? This will be discussed next.

REFERENCES

1. Stipp, D. (2006) Take two possibly lethal pills and call me in the morning. *Fortune*, February 20.
2. Singer, N. (2011) Ingredients of shady origins, posing as supplements. *New York Times*, August 27.
3. De Kosky, S. T., Williamson, J. D., Fitzpatrick, A. L., Kronmal, R. A., Ives, D. G., Saxton, J. A., Lopez, O. L., Burke, G., Carlson, M. C., Fried, L. P., Kuller, L. H., Robbins, J. A., Tracy, R. P., Woolard, N. F., Dunn, L., Snitz, B. E., Nahin, R. L., Furberg, C. D. (2008) *Ginkgo biloba* for prevention of dementia: A randomized controlled trial. *Journal of the American Medical Association*, **300**, 2253–2262.
4. Bent, S., Kane, C., Shinohara, K., Neuhaus, J., Hudes, E. S., Goldberg, D. O., Avins, A. L. (2006) Saw palmetto for benign prostatic hyperplasia. *New England Journal of Medicine*, **354**, 557–566.
5. Barry, M. J., Meleth, S., Lee, J. Y., Kreder, K. J., Avins, A. L., Nickel, J. C., Roehrborn, C. G., Crawford, E. D., Foster, H. E., Kaplan, S. A., McCullough, A., Andriole, G. L., Naslund, M. J., Williams, O. D., Kusek, J. W., Myers, C. M., Betz, J. M., Cantor, A., McVary, K. T. (2011) Effects of increasing doses of saw palmetto extract on lower urinary tract symptoms: A randomized trial. *Journal of the American Medical Association*, **306**, 1344–1351.
6. West, R., Zatonski, W., Cedeynska, M., Lewandoska, D., Pazik, J., Aveyard, P., Stapleton, J. (2011) Placebo-controlled trial of cytisine for smoking cessation. *New England Journal of Medicine*, **365**, 1193–1200.
7. Jorenby, D. E., Hays, J. T., Rigotti, N. A., Azoulay, S., Watsky, E. J., Williams, K. E., Billing, C. B., Gong, J., Reeves, K. R. (2006) Efficacy of varenicline, an $\alpha 4\beta 2$ nicotine acetylcholine receptor partial

agonist, *vs.* placebo or sustained-release bupropion for smoking cessation. A randomized controlled trial. *Journal of the American Medical Association*, **296**, 56–63.

8. Etter, J. F. (2009) Cytisine for smoking cessation. A literature review and a meta-analysis. *Archives of Internal Medicine*, **166**, 1553–1559.

9. Mann, J. J. (2005) The medical management of depression. *New England Journal of Medicine*, **353**, 1819–1834.

10. Walsh, B. T., Seidman, S. N., Sysko, R., Gould, M. (2002) Placebo response in studies of major depression. *Journal of the American Medical Association*, **287**, 1840–1847.

11. Wang, S. S. (2012) Why placebos work wonders. From weight loss to fertility, new legitimacy for "fake" treatments. *Wall Street Journal*, January 3.

12. Prayle, A. P., Hurley, M. N., Smith, A. R. (2012) Compliance with mandatory reporting of clinical trial results on ClinicalTrials.gov: Cross sectional study. *British Medical Journal*, **344**, d7373.

13. Ross., J. S., Tse, T., Zarin, D. A., Zhou, L., Kaumholz, H. M. (2012) Publication of NIH funded trials registered in ClinicalTrials.gov: Cross sectional analysis. *British Medical Journal*, **344**, d7292.

14. Appel, L. J., Clark. J. M., Yeh, H-C., Wang, N. Y., Coughlin, J. W., Drum, T. G., Miller, E. R., Dalcin, A., Jerome, G. I., Geller, S., Noronha, G., Pozefsky, T., Charleston, J., Reynolds, J. B., Durkin, N., Rubin, R. R., Louis, T. A., Brancati, F. L. (2011) Comparative effectiveness of weight-loss interventions in clinical practice. *New England Journal of Medicine*, **365**, 1959–1968.

15. Wadden, T. A., Volger, S., Sariner, D. B., Vetter, M. L., Tsai, A. G., Berkowitz, R. I., Kumanyika, S., Schmitz, K., Diewald, L. K., Barg, R., Chittams, J., Moore, R. H. (2011) A two-year randomized trial of obesity treatment in primary care practice. *New England Journal of Medicine*, **365**, 1969–1979.

16. Berry, J. D., Dyer, A., Cai, X., Garside, D. B., Ning, H., Thomas, A., Greenland, P., VanHorn, L., Tracy, R. P., Lloyd-Jones, D. M. (2012) Lifetime risks of cardiovascular disease. *New England Journal of Medicine* **366**, 321–330.

17. Heart Protection Study Collaborative Group (2001) Effects of 11-year mortality and morbidity of lowering LDL cholesterol with simvastatin for about 5 years in 20,536 high-risk individuals: A randomized controlled trial. *The Lancet*, **378**, 2013–2020.

18. The AIM-HIGH Investigators (2011) Niacin in patients with low HDL cholesterol levels receiving intensive statin therapy. *New England Journal of Medicine*, **365**, 2255–2267.

19. Moynihan, R., Heath, I., Henry, D. (2002) Selling sickness: The pharmaceutical industry and disease mongering. *British Medical Journal*, **324**, 886–891.

20. Brody, J. (2011) A reminder on maintaining bone health. *New York Times*, November 1.

21. Lenzer, R. (2011) Disease creep: How we're fooled into using more medicine than we need. Available at www.newamerica.net/node/61818.

CHAPTER 2

WHAT HAS HAPPENED TO R&D PRODUCTIVITY?

Simply stated, without a dramatic increase in R&D productivity, today's pharmaceutical industry cannot sustain sufficient innovation to replace the loss of revenues due to patent expirations for successful products.

—Steven Paul et al., *Nature Reviews Drug Discovery*, 2010

In both the scientific literature and the popular press, articles bemoan the perceived decreasing productivity of the R&D divisions of biopharmaceutical companies. It is not unusual to see titles like: "Is Pharma Running Out of Brainy Ideas?"[1] or "How to Improve R&D Productivity: The Pharmaceutical Industry's Grand Challenge"[2] These publications start from the premise that productivity is declining. Ironically, a 2005 paper by Schmid and Smith entitled "Is Declining Innovation in the Pharmaceutical Industry a Myth?" disputed this.[3] In their piece, Schmid and Smith looked at industry productivity in 10-year periods from 1945 to 2004 and essentially show that, by analyzing the data in this way, industry productivity has grown over the decades. As depicted in Figure 2.1, the number of chemical entities has steadily increased with the exception of the timeframe from 1965 to 1974. This decade was impacted by US legislation that overhauled the drug approval process—namely the Kefauver–Harris Drug Act, which revamped regulatory requirements for drug approvals. These more stringent requirements for New Drug Application (NDA) approvals reset the bar in terms of the need for more data and testing of experimental medicines at NDA filing. The productivity drop of 1965–1974 proved to be temporal, and NDA approvals grew over the next 30 years.

Does that mean that the current critics of industry productivity are wrong in their assessments? I don't think so. From 1991 to 2000, the FDA approved 315 new drugs. Yet, despite the progress in technologies used for drug discovery and development, as well as the wealth of information that has emanated from the Human Genome Project, which has helped better understand disease processes, only 231 new medicines (NDAs for small molecules and BLAs for therapeutic biologics) were approved from 2001 to 2010. More concerning is the fact that the R&D budgets for

Devalued and Distrusted: Can the Pharmaceutical Industry Restore Its Broken Image?
First Edition. John L. LaMattina.
© 2013 John Wiley & Sons, Inc. Published 2013 by John Wiley & Sons, Inc.

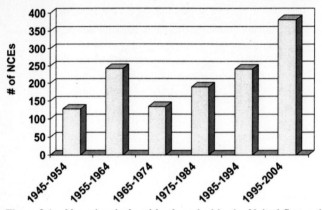

Figure 2.1   New chemical entities launched in the United States, 1945–2004.

biopharmaceutical companies in the last decade are higher than ever before. What is the cause of this drop-off?

In the last decade, the biopharmaceutical industry has been impacted in three major ways.

1. Industry consolidation
2. Heightened FDA requirements for NDAs
3. Higher hurdles set by payers

Each one of these factors has had a major impact on industry productivity in terms of the numbers of new drugs produced as well as the costs associated with making them successful products. Just how this has occurred and whether this situation is temporal (as it was in the 1965–1974 period) or has the potential to worsen even more will be discussed in the following section.

IMPACT OF MERGERS ON R&D PRODUCTIVITY

When people talk about the decline in R&D productivity, they always use 1996 as their starting point. In that year, the FDA approved 53 new medical entities (NMEs), an all-time high. However, using the 1996 data as a starting point for a productivity discussion is totally inappropriate because that was a one-year aberration. In the early 1990s, the United States was undergoing a "drug lag"; that is, drugs were being approved more rapidly abroad than in the United States. As a result, a number of drugs were languishing at the FDA for years before approval. Needless to say, Congress got involved and they realized that the FDA was under-resourced to approve drugs in a timely fashion. To solve this problem, Congress enacted the Prescription Drug User Fee Act (PDUFA), a mechanism whereby charges were levied on pharmaceutical companies for each new drug application filed. The revenues from these "user fees" were used to hire 600 new drug reviewers and support

Pharmacia & Upjohn
  Camptosar (irinotecan)
  Xalatan (latanoprost)
  Dostinex (cabergoline)
	Organon
	  Orgaran (danoparoid)
	  Remeron (mirtazapine)

Warner-Lambert
  Lipitor (atorvastatin)
  Cerebyx (fosphenytoin)
	Knoll
	  Mavick (trandolapril)

Rhone-Poulenc Rorer
  Zagam (sparfloxacin)
  Taxotere (docetaxel)
	Carter-Wallace
	  Astelin (azelastine)

Hoechst Marion Roussell
  Nilandiron (nilutamide)
  Allegra (fexofenadine)
	Athena Neurosciences
	  Zanaflex (tizanidine)

Dupont Merck
  Carbex (selegiline)
	Roberts
	  ProAmatine (midrodrine)

Figure 2.2   Some 1996 drug approvals from companies that no longer exist.

staff. This personnel increase enabled the FDA to work through the backlog of NDAs. The record number of NDA approvals in 1996 is a result of this.

(As an aside, the user fee in 1995 for a full NDA was $208,000. In 2012, the fee is $1,841,500. Considering that fewer NDAs are being filed and given the nine-fold increase in PDUFA user fees, one might wonder why all drugs can't be approved with a six-month review time.)

An underlying basis for the productivity of the 1990s was the large number of pharmaceutical companies that still operated 20 years ago. Many of the drugs approved in 1996 emanated from companies that no longer exist. A list of some such drugs appears in Figure 2.2. Interestingly, many of these compounds have become very important medicines to treat a variety of diseases, including cancer and heart disease.

It is striking to note how extensive industry consolidation has been. Figure 2.3a shows the members of the Pharmaceutical Research and Manufacturing Association (PhRMA) in 1988. Even veterans of the industry are hard-pressed to remember all of these organizations. Figure 2.3b shows those members of PhRMA that remain from 1988. The fact that 31 out of 42 were eliminated is stunning.

If you were to go back and read the corporate explanations for these mergers, they would make perfect business sense. In some cases, the product lines of the companies aligned closely, making the new company stronger from a commercial perspective. Yet another example was one in which a large company bought a smaller one to allow time for the larger company's pipeline to mature.

However, while each specific merger likely had immediate effects on the individual companies involved, the broad consolidation depicted in Figure 2.3b had a major impact on the overall productivity of the industry. The R&D portfolios of the original 42 companies, while differing in size, tended to be broader in scope than those of the companies that remain. It is likely that when a new idea for treating

Abbott Laboratories	G.D. Searle	Procter & Gamble
American Cyanamid	Glaxo	Rhone Poulenc
A.H. Robins	Hoechst	Rorer
Astra	Hoffmann-LaRoche	R.P. Scherer
BASF	ICI	Roussel
Beecham Laboratories	Johnson & Johnson	Sandoz
Boehringer Ingelheim	Knoll	Schering Plough
Boots Pharmaceuticals	Eli Lilly	SmithKline
Bristol-Myers	Marion Laboratories	Squibb
Carter-Wallace	Merck	Sterling Drug
Ciba Geigy	Merrell Dow	Upjohn Company
Connaught Laboratories	Monsanto	Warner-Lambert
DuPont Pharmaceuticals	Pfizer	Wellcome
Fisons	Pharmacia	Zeneca

Figure 2.3a   Members of the Pharmaceutical Research and Manufacturing Association (PhRMA) in 1988.

Abbott Laboratories	Eli Lilly
AstraZeneca	Merck
Boehringer Ingelheim	Novartis
Bristol-Myers Squibb	Pfizer
Glaxo SmithKline	Sanofi-Aventis
Johnson & Johnson	

Figure 2.3b   PhRMA members that remain from 1988.

cancer arose in 1990, 20 companies would have jumped on it. Given the difficulties encountered in R&D, one might assume that only three or four of these companies would be successful in bringing a drug based on this idea to market. Things are completely different now because far fewer competitors exist. With fewer companies now involved, the chances for success drop precipitously. This may explain why there are now significantly fewer entrants in a new clinical class than in the late 1990s.

One can look at the reduction in the number of pharmaceutical companies with a degree of nonchalance. After all, these mergers simply are the combination of two entities to form a much larger one. Furthermore, consolidations make great business sense. They remove duplication, reduce costs, and produce synergies. To a certain extent, this is all true. As was stated, in the early days of these mergers, organizations described these mergers as being part of a growth story with the ultimate goal of achieving greater scale. In these situations, for example with the merger of Bristol

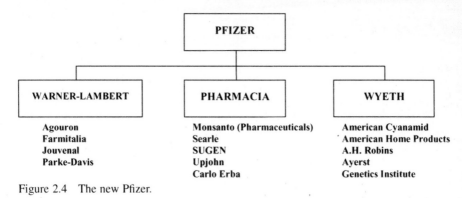

Figure 2.4   The new Pfizer.

Myers with Squibb, the R&D divisions were, in a sense, fused. Program overlap was minimized and new projects were added. Major R&D cuts didn't occur.

This has changed radically in the last decade. The term "synergy" is a euphemism for downsizing. In major mergers today, not only are R&D cuts made, but entire research sites are eliminated. The Pfizer experience is particularly telling. Prior to 1999, Pfizer had never done a major acquisition. Over the next decade, it accumulated Warner–Lambert, Pharmacia, and Wyeth (Figure 2.4), and each of these companies was the amalgamation of others. This does not include smaller biotech acquisitions like Vicuron, Rinat, and Esperion. Over this timeframe, in order to meet its business objectives, Pfizer has had to close or scale back numerous research sites, including major ones in Kalamazoo, Michigan (Upjohn), Ann Arbor, Michigan (Warner-Lambert), Skokie, Illinois (Searle), and Sandwich, UK (Pfizer). These are sites that housed thousands of scientists, sites where many major drugs including Xanax, Lipitor, Norvasc, and Viagra were discovered.

But Pfizer is not unique in this regard. Sanofi-Aventis has its roots in Marion Merrill Dow, Roussel-Uclaf, Hoechst, Rorer, Rhone-Poulenc, Fisons, Synthelabo, and Sanofi. Many of the research sites of these former companies have been abandoned. Even Merck, which for years had avoided major acquisitions, merged with Schering-Plough and thereby closed the Organon (Netherlands) and Merck-Frosst (Montreal) sites.[4] The latter housed the team that discovered Merck's biggest selling drug, Singulair.

However, there is another chilling aspect to such cuts. Historically, the pharmaceutical industry has prided itself on the fact that it invested more in R&D (as a percentage of revenues) than did any other industry. At times, companies have invested as much as 20% of top line revenues into their pipeline. Pfizer now projects that in 2012 this will only be 11%. This is not a one-year aberration. Forbes has analyzed that Pfizer will still be investing at this level in 2015.[5]

Equally concerning are the deep cuts being made in R&D budgets. Most dramatic is Pfizer's plan. In their Annual Report for 2008, Pfizer reported R&D expenses of $7.95 billion, and that for Wyeth was $3.37 billion ($11.3 billion total). Pfizer has announced that its projected 2012 R&D budget will be between $6.5 and $7 billion.[6] The magnitude of this drop is stunning; in effect, the equivalent of the entire Wyeth budget plus another billion dollars of investment in R&D has been eliminated.

Pfizer has not been alone. Other large pharmaceutical companies such as BMS, Abbott, and AstraZeneca have announced similar cuts. It remains to be seen whether all companies will regress to the degree Pfizer has. Furthermore, Biotech is not immune to this trend. Alnylam announced a 25–30% reduction in its workforce in 2010 and another 30% cut in 2012. But more pertinent to this sector is the report that 100 publicly traded biotech companies in the United States have been acquired or ceased operations in the last 3 years, a 25% drop overall.[7] These two trends may be related. Smaller biotech companies and particularly start-up companies look to Big Pharma for support of new drug programs, especially as they enter more expensive clinical trials. As Big Pharma R&D budgets shrink, there is less money to invest both internally and externally.

Thus, over the past 15 years, private sector R&D has been shrinking predominately in researchers but more recently in dollars invested. This is now occurring across the entire R&D health-care sector. Ironically, at a time when more is understood about the basis of various diseases, the ability to exploit this information in the private sector is being compromised. It is hard to envision that R&D productivity, as measured by new drug approvals, will improve in the coming years based on this reduced investment. This impact will not necessarily be felt immediately but rather in 5–8 years from now, given the delayed effect that cuts in research have on long-term portfolio prospects.

Another impact that mergers have is on the newly formed company's pipeline. In effect, the progress of compounds through the R&D continuum appears to slow as the result of a merger. An analysis of the AstraZeneca acquisition of Medimmune is revealing.[8] At the time this deal was consummated, Medimmune had a pipeline of 12 compounds: one in Phase 3, two in Phase 2, and nine in Phase 1. Two years later, the Phase 3 compound was filed, two Phase 1 compounds moved into Phase 2, and three compounds were dropped. The remaining six drugs' status was unchanged. These results would certainly be on the low end of project advancement in the industry.

More telling is the situation for Pfizer. For the past few years, Pfizer has updated its pipeline every six months on its website, so all can track the progress of its clinical programs. In comparing the February 2008 data with February 2011 data, one can also see a relatively slow pace for a number of compounds. As an example, looking at the Pfizer Phase 2 portfolio, not including the compounds that came from Wyeth, 40% of the compounds have been in Phase 2 for more than three years. Again, this is below industry average.[9]

Mergers and reorganizations take a toll on any organization, but R&D divisions seem to be especially vulnerable.[10] Having a sense of how merging R&D organizations occurs is helpful to understand this. First of all, the R&D organizations of the two organizations are the last to begin merger discussions before formal approval by the Federal Trade Commission (FTC). The reason for this is that a company's pipeline and intellectual property are considered their "crown jewels." Exposing either side to an in-depth review of each other's data would be lethal if the proposed merger were to derail for any reason. Thus, the R&D leaders of the respective organizations do not meet until late in the process, later than colleagues in other parts of the company.

When the R&D discussions finally occur, the initial focus is on late-stage Phase 3 programs, followed by reviews of the mid-stage candidates. The early discovery programs are handled last. These reviews are extensive and time-consuming. Clinical candidates get scrutinized in great depth: Compound efficacy results, emerging adverse event profile, mode of production, and costs of goods are just some of the analyses done. In addition, commercial evaluations are made for each compound.

Depending on the type of merger, the goals of these discussions vary. In a "fusion merger" where both organizations are being united for growth, the main objectives of these discussions are to identify areas of overlap, eliminate duplication, identify any anti-trust issues, understand portfolio or compound specific issues, and determine the new structure of the organization. In addition, research organizations often differ procedurally in some fundamental processes like IT platforms, data handling, or adverse event monitoring. Sorting through which system to use or to create a hybrid takes time not only for selection purposes but also for conversion to the new adopted process.

In a "synergy merger," one in which extensive cost cutting occurs, these discussions have a much harder edge. Not only is the goal to eliminate duplicate programs, but also information is gathered as to which research sites can be eliminated, what therapeutic areas of research can be exited, which clinical programs are not a strategic fit for the new company, who will lead the new organization, and so on.

Against this backdrop, one can understand how things can slow down. For example, during this period which can take at least nine months, generally no new programs are started. This makes sense in that if the company has set cost targets for R&D for the new organization, one does not want to start new studies in programs that might be dropped six months later. Hiring freezes also occur, because it does not make sense to hire a new scientist to run a program that might not be continued. In areas of overlap, decisions need to be made as to which program should be continued and which should be dropped. All of these factors result in the dampening of any momentum that might have existed before the proposed merger was announced. In effect, many things essentially go on hold for this period. For a company to undergo such a scenario once has a significant impact on timing. For a company to endure this multiple times could prove crippling.

Beyond the impacts on productivity and speed of execution, mergers have social consequences on employees.

Imagine waking up one morning and hearing that your company is about to merge with another. You would probably read the announcement with interest and try to grasp not only the reasons for the merger but also the impact on yourself. A variety of questions are likely to arise immediately but, in particular, there are some of key interest to you. You will worry:

1. Whether your research site will survive the merger
2. Whether your project will continue

3. Whether your position will be retained or eliminated

4. Whether you'll be required to move to a new location to keep your job

If you are in a leadership role, you will have similar questions, but you will also realize that there is likely to be a leader in the other organization who has the same job. In fact, there are probably multiple candidates for your job. Noncollaborative behaviors begin to set in.

When my views of the impacts of mergers on R&D first appeared,[11] they generated a lot of comments from many people, mostly scientists, about how they were affected. One critic of mergers is Dr. Raymond Firestone, an eminent researcher in drug discovery. In a recent article entitled "Lessons from 54 Years of Research," his account of the 1989 merger of Bristol Myers and Squibb is quite telling.[12]

> If size is detrimental to an innovative research culture, mergers between large companies should make things worse—and they do. They have a strong negative personal impact on researchers and, consequently, the innovative research environment. For example, the merger of Bristol Myers with Squibb in 1989, which I witnessed, was a scene of power grabs and disintegrating morale. Researchers who could get a good offer left the company, and the positions of those who remained were often decided by favouritism rather than talent. Productivity fell so low that an outside firm was hired to find out why. Of course, everyone knew what was wrong but few—if any—had the nerve to say it.

Another veteran of this same merger told me the following:

> I had been at Squibb for 15 years and was right there. You're right—it was prettier than some that came after, but it was still very ugly. Good people got thrown out because they were redundant.

> Perhaps the best measure of that merger was to answer the question: "How long did it take before a true BM-S drug emerged from the new company?" I think we are talking at least ten years.

This is obviously a trying time for all concerned, and things are worsened by the fact that it may be a year before you know the answers to all of these questions.

It is hard to quantify the impact on individual productivity caused by the announcement of a merger or acquisition. As months of discussions and decision making go by, rumors are rampant. Various websites such as Biofind or Cafepharma allow rumors to spread instantaneously. On any day, one can go on any such website and know which companies are in turmoil by the number of posts and the type of comments that appear. It is hard to focus on your project if both your status and your project are in doubt. Furthermore, it is emotionally draining. One scientist, whose entire six-and-a-half-year tenure was under the cloud of rumored layoffs, finally got axed. He said that while he was initially upset, he "was really relieved; the stress at work was incredible."[13]

During this timeframe, an organization is ripe for recruiters. Given your personal uncertainty, not just in job potential but also location, it is likely that you would be attracted to a potential position at another firm. Not just focus, but also one's commitment, gets severely tested in these times.

It's been said by leaders of organizations who have done multiple mergers that "We've done this before and we know how to do it." Mechanically, that is certainly true. However, the organization has also survived it, and the memories are not fond ones. The thought of repeating the exercise, particularly when one's own future is uncertain, is not embraced. It could, in fact, prove numbing.

Independent of mergers, when cuts are planned for an R&D organization, if not done thoughtfully, morale can also suffer. In early 2012, AstraZeneca announced that it would be making layoffs in its organization. Derek Lowe, one of the best bloggers in the biopharmaceutical industry, posted these comments on February 3, in his "In the Pipeline." They capture this type of atmosphere very well.

> From several reports, here's what I have on AstraZeneca's plans in Waltham: they've told people there that cuts are coming. But they haven't gotten very specific on when, or who, or how many. All those questions (that is, all the questions there could be) are under review.
>
> Pfizer has done this to their people before, as have other companies in the throes of layoffs, and it's the only way I know to actually push morale and productivity *down even further* in such a situation. You come to work for weeks, for months, not knowing if your project, your lab, or your whole department is heading for the chopping block. All you're sure of is that someone is. And will your own stellar performance persuade upper management to keep you, when the time comes? Not likely, under these conditions—it'll more likely be the sort of thing where they draw lines through whole areas. Your fate, most people feel at these times, is not in your own hands. A less motivating environment couldn't be engineered on purpose.
>
> But that's what AZ's management has chosen to do at their largest research site in North America. I hope that they enjoy the results. But then (and more on this later), these are the people who have chosen to spend billions buying back their own stock rather than put it into research in the first place. It's not like the score isn't already up there on the big screen for everyone to see.

Despite these obvious downsides, it is unlikely that the era of large mergers in the pharmaceutical industry has ended. The Chief Strategy Officer at Glaxo, David Redfern, essentially said as much at a recent meeting[14]: "As the major western markets slow down, M&A has been seen as a way of getting more top-line growth." The issue with this strategy is that, although you are growing your top-line significantly in a merger, your R&D engine is not growing robustly enough to keep pace. Thus, the cycle of more M&A to fuel corporate growth continues.

Not every major pharmaceutical company CEO believes in this strategy. Lilly chief, John Lechleiter, himself a scientist, has publicly stated that: "We are still very much opposed to a large-scale combination. We don't think size is necessarily supportive of innovation." And not all CEOs of merged companies believe in cutting back R&D. Merck's newly appointed CEO, Kenneth Frazier, has recently stated Merck will focus on investing in drug development to drive growth.[15] In fact, he has sought to reinvest some of the savings realized in the Schering-Plough merger into Merck's promising but costly late-stage pipeline.

Whether others in the pharmaceutical industry come around to the views of Lechleiter and Frazier remains to be seen. However, evidence from the last decade should give pharmaceutical industry leaders' pause when considering major mergers. The impact of such deals on an internal R&D organization is dramatic from productivity and morale perspectives.

As people hear my views on this topic, they show a good deal of skepticism. After all, I had a leadership role at Pfizer during two large mergers: the first with Warner-Lambert and the second with Pharmacia. I can't blame the skeptics for questioning my current position. However, after having lived through these mergers and having dealt with cost-cutting, site closures, shifting of people and projects, and so on, I have come to the realization that, on balance, these mergers hurt organizations more than they help.

But mergers in the pharmaceutical industry should not just concern those who work in these R&D groups. Such mergers should be equally concerning to patients, physicians, and payers. Recent Big Pharma cutbacks in areas of research such as antibacterials and neurosciences have raised significant concerns for those who treat these diseases.[16] Industry consolidation of the last 15 years has resulted in less competition, less investment in R&D, and a gradual decrease in the approval of new medicines. At a time when the world is still in dire need of new treatments for a host of conditions such as Alzheimer's disease, drug-resistant infections, diabetes, and so on, such a trend should be alarming.

HEIGHTENED FDA REQUIREMENTS FOR NDAs

> There were variable readings of the FDA in different quarters of American industrial society, images of the regulator that ranged from cooperative enforcer to vigilant purity cop to industry bully to vanguard assessor to bought-off corporate servant.

This quote could be appropriate in describing the FDA today. Actually, it is a quote from Daniel Carpenter in his book, *Reputation and Power: Organizational Image and Pharmaceutical Regulation at the FDA*,[17] and he was referring to how the FDA was viewed in the 1940s. One might argue that the agency is viewed in the same way today. There are those who believe that the FDA is too conservative, that they take too long to approve lifesaving medicines, or that they erect hurdles so high that they discourage companies from even seeking new medicines. Yet, there are probably an equal number of FDA critics who believe that the FDA is too cozy with the biopharmaceutical industry and that they approve drugs without having enough data to justify exposing the broad population to such drugs.

On August 10, 2011, the *Wall Street Journal* published a letter to the editor from Dr. John Freund, the founder of Skyline Ventures, a capital firm specializing in biopharmaceutical drug companies. In his letter, Freund lashed out at the FDA. Specifically, he challenged the view taken by Dr. Margaret Hamburg, FDA Commissioner, in an Op-Ed piece she wrote in the *Journal* entitled "America's Innovative Agency: The FDA."[18] In her piece, Hamburg stated that:

> When presented with high-quality applications that are based on strong science, we work quickly and efficiently. And we must, because our mission is to promote and protect the American people.

The FDA's self-anointed role as "Innovative Agency" incensed Freund, who wrote:

> The FDA has tightened up the requirements for approving new drugs for adult-onset diabetes, a disease that affects approximately 25 million Americans. The result is that performing the clinical trials for a new diabetes drug is so long and costly that no venture capital firm will finance a new diabetes drug.

He also had doubts about the FDA's desire to approve new drugs to treat obesity, writing:

> In the past year, the FDA rejected three separate drugs to treat obesity, including one for which the FDA's own advisory panel recommended approval. As a result, no venture capital firm will now finance a new effort to develop a drug for the obesity epidemic.

Freund is not alone in these views. The CEO of the French pharmaceutical company Sanofi, Chris Viehbacher, echoed these same sentiments. Viehbacher, who also serves as the chairman of the board of the Pharmaceutical Research and Manufacturers Association (PhRMA), stated that:

> You're starting to see primary care diseases becoming somewhat neglected.[19]

His concern is that companies cannot predict how the FDA will weigh the risk–benefit for drugs to treat diabetes and obesity. The result is that companies are shying away from R&D in this area.

In reading this, one would think that the FDA is way too conservative in evaluating drugs, particularly in treating chronic diseases like obesity and diabetes that affect millions of Americans. But then, one must factor in the case with Avandia. Discovered and marketed by Glaxo (GSK), Avandia is a treatment that helps diabetics to lower their blood sugar. The ability of a drug to lower blood sugar had been the medical standard used by regulators for decades to approve new drugs for this disease. Avandia was a highly successful product for GSK, and sales exceeded $3 billion annually. However, over time it became clear that, although Avandia helped control a patient's blood sugar, its usage appeared to elevate the risk of heart attacks and strokes. This was counterintuitive. Diabetes, if left untreated, can result in heart attacks and strokes. Inexplicably, Avandia wasn't providing the long-term benefit of reduced cardiovascular risk despite lower blood sugar.

Needless to say, the FDA came under harsh criticism as the Avandia situation unfolded. Here is a typical response reported by *Time* magazine in an article entitled "After Avandia, Does the FDA Have a Drug Problem?"[20]

> Over the past two decades, as drug after drug has been recalled after winning FDA approval, it has been hard not to wonder if FDA regulators have been captured by the drug industry. FDA critics and industry monitors charge that the drug-approval process is too easy for pharmaceutical companies to game. It is in some ways an unsurprising development. The FDA serves a public insatiably hungry for new medicines. Yet the agency does not have responsibility for performing safety testing. It relies on drug

companies to perform all premarket testing on drugs for safety and efficacy. And it relies on industry "user fees" for 65% of its budget for postmarket monitoring of the drugs it approves, thanks to a 1992 law designed to speed treatment to patients. 'The FDA's relationship with the drug industry (is) too cozy', says Senator Chuck Grassley of Iowa.

By late 2010, the FDA greatly restricted the use of Avandia, essentially limiting its use to special circumstances. Regulatory agencies around the world did the same. As a result, GSK reported that in 2011, global Avandia sales were less than $190 million.

But, the more important outcome from the Avandia situation and similar events is that the FDA is now requiring more data before approving drugs to treat diabetes and obesity.

This relatively conservative position taken by the FDA in obesity and diabetes is not unique to these therapeutic areas. In fact, it is seen in new drugs for cancer, osteoporosis, and heart disease. No longer will the FDA approve a drug for obesity based solely on its ability to induce weight loss. Similarly, the FDA wants to see more than blood sugar lowering before approving a drug for diabetes. While both are meaningful markers for improving the respective disease condition, the FDA also wants to see outcome studies—that is, 2- to 3-year-long studies showing that the weight loss or lowering of blood sugar actually correlates to a reduction in heart attacks and strokes, the unfortunate end results for diabetes/obesity.

Why is the FDA asking for long-term data? Won't weight loss and/or blood sugar lowering automatically result in enhanced survival? Based on the Avandia experience, the answer is no.

The same is true for new drugs in other classes. Studies have shown that just because a drug shrinks the size of a tumor doesn't mean it enhances survival. Or, if a compound stabilizes bone formation, it may not prevent fractures. And there are a number of studies that show that raising HDL, the good cholesterol, doesn't reduce heart attacks. Faced with this situation, the FDA has done the logical thing: It wants outcome studies to justify approval for drugs that will need to be used by patients for decades.

As Dr. Freund pointed out, the need for new obesity drugs is a particular challenge for the FDA. The need for such a medication is obvious. With more than 30% of the American adult population now characterized as obese, more than diet and exercise is needed to get this problem under control. But, here's the problem. The FDA is well aware that any drug that they approve for obesity would be immediately used by millions of people. Such a medicine would not be used for just the morbidly obese. People anxious to lose a few pounds for cosmetic reasons would clamor for it. The FDA is acutely aware of this and so is wary of unleashing such a drug without *extensive* safety and efficacy data.

It is hard to criticize the FDA for taking such a position. Diet drugs have a history of having modest efficacy but significant side effects. Most notorious was a drug known as fen-phen, which was pulled from the market in 1997 as a result of valvular heart disease. Thus, the FDA has taken the following stance: Any drug for obesity must not only show efficacy, but must also be studied in patients over the course of 2–3 years to (1) demonstrate long-term efficacy in weight loss and (2) show long-term safety in patients who will take this drug over the course of many years. On this latter point, the FDA wants to see outcome studies—studies in which patients on the drug are followed for multiple years to see what the impact of the drug is on heart attacks and strokes. Such a request is not unreasonable. After all, a drug that causes weight loss should, at the very least, not increase heart attacks and, in theory, should cause a reduction of heart disease.

The requirement of such an outcomes study has caused the biopharmaceutical industry to shy away from this disease area. The reason is simple: You can invest between $500 million to $1 billion dollars in an obesity R&D program over 10 or more years only to learn at the last step, the outcome study, that your drug is not approvable. That is a level of risk that is not acceptable to investors like Freund.

This whole topic has been recently revisited with a new drug for obesity called Contrave. This drug is being developed by a small biotech company, Orexigen. Contrave is actually a combination of two drugs that have been on the market for decades: naltrexone (used to treat alcohol addiction) and bupropion (a smoking cessation medication initially discovered as an antidepressant). The scientific rationale for using this combination is as follows. Naltrexone blocks opioid receptors in the brain, reduces the reinforcing aspects of addictive substances, and negatively alters the taste of many foods, including sweets. Bupropion increases dopamine activity in the brain, and this appears to lead to appetite suppression and increased energy expenditure.

This sounds pretty exciting. Orexigen is utilizing two marketed agents and combining them into a single pill that dulls your appetite, reduces your craving for sweets, and causes your body to expend more energy. More importantly, clinical trials show that Contrave actually works. Patients on the drug for 56 weeks lose an average of 6% of body fat. This is not dramatic weight loss, but it can help in one's overall weight loss program.

Nevertheless, the FDA initially rejected this drug. Despite the fact that Contrave is the combination of two known drugs, the FDA wants to see the results of a long-term cardiovascular outcomes study before approval. Orexigen recently published the results of their meetings with the FDA, which outlined the final regulatory requirements for Contrave approval. Essentially, the FDA has asked for a trial comparing Contrave to placebo in a population of overweight and obese patients who have an estimated background rate of 1–1.5% annual risk of major cardiovascular events. The length of this study is dependent upon seeing a specific number of cardiovascular events (heart attacks, strokes, revascularizations, etc.) for the FDA to be able to judge whether Contrave poses no risk in this population.

Thus, the road to approval is clear for Orexigen. However, it is not without risk. First of all, this study will require 10,000 patients. If one assumes a cost of $10,000 per patient, this will require an investment of at least $100,000,000. Second, while the FDA might approve this drug if the study finds that patients lose 6% of body fat with no increase in major adverse cardiovascular events, will such a modest decrease in weight result in insurance companies paying for it? Payers might take the stance that this is a "bikini drug": It causes you to lose some weight, but there is no long-term health benefit from such weight loss. They could deem Contrave to be a "cosmetic drug" and not a real medicine. Thus, patients who want Contrave would have to pay for it themselves—which they might do.

Of course, Orexigen could hit a home run with the following result: Significant weight loss is seen, the drug is well-tolerated, and, most importantly, a significant drop in adverse cardiovascular results is seen as a result of this medicine. Such is the nature of pharmaceutical R&D—high risk, but the potential for high rewards for the patient and the drug company.

Dr. Freund's assertion that "no venture capital firm will now finance a new effort to develop a drug" for diabetes and obesity is not true. Catabasis, a biotech company in Cambridge, MA, recently raised almost $30 million from VCs to support its exciting new approach to diabetes. And Gelesis, a company that has been formed by PureTech Ventures (where I am a Senior Partner), is focused on obesity and has also recently raised funds to support its R&D program. Despite this admittedly tough regulatory environment, good ideas are still garnering investments.

There is a positive aspect to all of this. If a compound can successfully clear these hurdles, it will be a blockbuster. The manufacturer will be able to say that its new compound not only causes weight loss, but also reduces heart attacks and strokes. The same would be true for a diabetes drug that reduces cardiovascular events as well.

In effect, one could argue that the FDA has evolved from approving drugs that address symptoms to seeking treatments that have a fundamental impact on disease. Generating such data is costly—but essential if a company hopes to be successful in the field of chronic diseases.

Not too surprisingly, there are those who argue for fundamental changes to the FDA's mandate. An opinion piece in the *Wall Street Journal* by Michele Boldrin and S. Joshua Swamidass called "A New Bargain for Drug Approvals," lobbies for a fundamental change in FDA[21] oversight for new drug approvals. Basically, Boldrin and Swamidass are advocating for a system that puts safety first and allows for the proof of effectiveness later. Specifically, the authors argue that the FDA "should return to its earlier mission of ensuring safety and leaving proof of efficacy for post-approval studies and surveillance. It is ensuring the efficacy—not the safety—of drugs that is most expensive, time-consuming and difficult."

What is their reason for proposing such a change? Among other things, they are concerned that the necessity of proving a drug's efficacy, which is the primary driver for the approximately $1 billion price tag in developing a drug, is limiting the number of compounds that the biopharmaceutical industry can advance through FDA approval. The authors believe that if a company only had to invest in showing that a compound was safe, the entire process would be cheaper and then companies would have the funds to "unleash the next wave of medical innovations."

They go on to propose that "In exchange for this simplification, companies would sell medications at a regulated price equal to the total economic cost until proven effective, after which the FDA would allow the medications to be sold at market prices."

There are a variety of obstacles that such a proposal would face. First of all, there is a reason that the FDA's remit was expanded decades ago. In evaluating a medicine, you need to be able to put the safety profile of a new drug in perspective with its benefit profile. A breakthrough cancer drug with some serious side effects might be justifiably approved because of its lifesaving effects, whereas a compound that lowered LDL but which has this same side-effect profile might be deemed unapprovable based on the benefit–risk profile. You need to have full safety and efficacy data to judge a new drug fully.

The second major issue is economics. Drug prices are already regulated throughout the world. Unlike the situation with most products, there is no "market price." Instead, in most countries the price of a drug is set by the government. It's likely that price will also be more regulated in the United States, where the government is becoming the major consumer. In addition, payers (HMOs, etc.) want to know the safety and efficacy benefits of a new drug before they reimburse prescription costs. Thus, having efficacy data is needed to help get a new drug prescribed as well as to get a reasonable price. Without having full efficacy data, the drug will not be made broadly available to patients.

Finally, and maybe most importantly, why would a physician prescribe a new drug without having a full understanding of the beneficial effects of such a compound as well as knowing how the new drug stacks up in comparison to the treatments that she/he is already using? Just because it may be deemed safe isn't a good enough reason for a physician to give it to patients. Even when you have these data now, physicians are reluctant to try a new medicine if they already have found good success with an established one.

Call me old-fashioned, but I like having the FDA as the independent evaluator of both safety and efficacy. Such approval gives doctors and patients confidence in the health-care system. Abandoning this practice would be a big mistake.

The FDA, while not perfect, is often criticized unfairly. Here are three typical examples. Invariably, after a new drug is on the market for a few years, the FDA will announce that this new medicine has some new safety issue that hadn't been seen previously. This issue is not unusual. When a new drug is approved, it generally has been tested in perhaps 10,000 patients for a year. Once it is on the market, if it is a drug for chronic therapy, it may get used by millions of people over the period of a few years. As it is more broadly used, more information is gathered on the drug and the FDA will react to this. Unfortunately, when this occurs, there will be members of Congress (Senator Grassley comes to mind) who will hold a press conference and rail against the FDA for its lax attitude in approving new drugs. There are *always* risk–benefit issues with any medicine, old or new. Some politicians don't seem to get this, or pretend that they don't. The FDA does.

There are those who criticize the FDA for being too slow in approving new drugs. This can occur in diseases where there is no adequate treatment for life-threatening diseases. In these cases, there are patient advocacy groups who want certain medicines approved as quickly as possible. Their view is that, if there is a drug in trials that can save their lives or those of their loved ones, the FDA has no right to prevent patients from getting this medicine. In these cases, the FDA generally has significant questions about whether the drug indeed is effective and will not approve it without hard data showing that the drug works. It should be noted that the FDA has a good track record of reacting with alacrity to health crises, as evidenced by its actions in approving HIV drugs in the early days of the AIDS epidemic and, more recently, in approving new cancer drugs.

But what is most exasperating is the criticism that the FDA is beholden to the pharmaceutical industry because of the Prescription Drug User Fee Act (PDUFA). Congress enacted PDUFA, as a charge to the pharmaceutical companies on the filing of their New Drug Application (NDA). The aim was to help provide resources the FDA needed to review NDAs in a timely fashion. Critics say that, because the FDA is dependent on these user fees to support itself, FDA staffers feel beholden to pharmaceutical companies and, as a result, try to do all they can to help these companies. This view is ludicrous. The fee is paid whether the FDA approves or rejects the NDA.

To a certain extent, the FDA is in a no-win position. It has to judge what medicines the US population should or shouldn't take. This is rarely a black or white decision. As was stated above, all medicines have side effects. The question the FDA must answer is what the risk–benefit of each new drug is—this is always a judgment call. The FDA tries to work closely with all of those dependent on its actions: patients, physicians, pharma companies, payers, and government officials. The FDA is not infallible, but it gets it right most of the time.

So, how do the increased requirements for new drug applications, as set by the FDA, impact overall industry productivity? For one thing, costs for developing a new drug escalate greatly as a result of the need for outcome studies. These types of studies tend to require large numbers of patients, anywhere from 10,000 to 30,000, depending on the nature of the study. Plus, they generally need to run for three to five years because the outcomes normally are medical events that take years to develop. This combination of patient numbers and time drive the overall costs of such a study to be anywhere from $100 and $300 million. If you have multiple such programs in your development portfolio, then a sizable chunk of your R&D budget is consumed by these studies, thereby taking away from other projects. Thus, overall portfolio yield is down.

A second impact of these requirements is that the development timeline for a new drug in the chronic disease area is lengthened as a result of the need to do these studies. This has impacted overall industry productivity in the last decade. Compounds that might have been launched a few years ago may now be undergoing 3- to 5-year long-term studies. Theoretically, this impact may be temporary. Once this delay is built into a company's timeline for its pipeline, the overall productivity could again turn upward. But, for drugs to treat chronic diseases, the time from having an idea to convert that idea to an approved new medicine will be 15 years rather than 10.

Finally, in the 1990s, when the FDA approved drugs based on treating symptoms like pain relief, high cholesterol, or high blood glucose, outcome studies would be done after approval—if at all. A drug like Avandia and others in its class, such as Rezulin or Actos, would not be approved today. Had outcome studies been done with any of these compounds before approval, the adverse cardiovascular effects would have been found prior to launch. There are compounds approved in the 1990s that would not be approved today. Thus, the type of industry productivity seen in the 1990s may never be repeated.

There is no doubt that these changes have impacted the industry. But there are two positives to this. First, any compound that can successfully wend its way through these higher hurdles should be certain of success. Having been tested in such a setting and showing its value in impacting the disease, such a drug would be in demand by patients and physicians. Furthermore, payers would be hard pressed to deny coverage of such a drug. But perhaps the bigger positive would come from the fact that the greater amount of safety data gathered on the drug before it is marketed could provide some degree of comfort that the FDA won't be called to remove the drug after its launch because of an unfavorable risk–benefit profile.

HIGHER HURDLES SET BY PAYERS

The headlines were pretty impressive:

> Scientists crack histamine code that could lead to better allergy relief.

These stories were based on an article in the scientific journal *Nature*, describing the work of an international research team.[22] This group studied how doxepin, an old drug used to treat allergic reactions and other conditions, binds to the human histamine receptor protein, thereby easing allergy symptoms. The team did this by solving the complex three-dimensional structure of the human histamine H_1 receptor protein. Dr. Simone Weyland, one of the investigators, commented on the importance of this finding:

> First-generation antihistamines such as doxepin are effective, but not very selective, and because of penetration across the blood–brain barrier, they can cause side effects such as sedation, dry mouth, and arrhythmia. By showing exactly how histamines bind to the H_1 receptor at the molecular level, we can design and develop much more targeted treatments.

Based on this pronouncement, allergy sufferers might assume that they will someday have new, superior medicines to treat their symptoms. Unfortunately, this scenario is highly unlikely.

Antihistamines, like diphenhydramine (the active ingredient of Benadryl), have been available for over 60 years. While effective in treating hives, runny nose, itchy eyes, and so on, they do cause toleration issues, including sleepiness. In the 1980s, pharmaceutical companies sought to come up with better antihistamines that produced less drowsiness. This pursuit led to the so-called second-generation agents such as loratidine (brand name, Claritin) and cetirizine (brand name, Zyrtec). These antihistamines have proven to be safe; they are so broadly used that they are now available over-the-counter at any pharmacy and supermarket.

Perhaps loratidine and cetirizine can be improved upon. The research that's been done to get a better understanding of the biological interaction between the drug and the site of action in the body would be a good starting point to begin such a drug discovery program. But it would be surprising to find a company willing to make such an investment. The current antihistamines on the market work well enough. To do the research to devise a new generation of agents from lab bench through clinical trials and to FDA approval would require 12–15 years and, at the very least, hundreds of millions of dollars.

But, even if you made this investment of time and resources, and even if you had an antihistamine that was better tolerated than current agents and even if the FDA approved your NDA, you still have a big problem—payers. Why would an insurance company be willing to reimburse payments for a premium-priced new drug with modest advantages over cheap generic or over-the-counter medications?

In the 1990s, companies would have eagerly begun a race for a third-generation antihistamine. This doesn't happen anymore. The return on investment (ROI) for such a drug would be negative. It would be economically irresponsible for a company to undertake such a program.

To help decide whether research programs are worthy of pursuit, companies have established pharmacoeconomics groups. These teams analyze the value of the potential new drug as compared to existing therapy. This value can be expressed in a variety of ways: relative effectiveness of the new drug, enhancement of the patients' quality of life, and the reduced burden of costs to society that the drug can bring. The goal of this work is to assure companies that, if they successfully navigate the drug approval process, the global health-care system will pay for it.

In an interview with Reuters,[23] Shire CEO, Angus Russell, captured things this way:

> If you're going to go out with a drug that you don't know whether it's better than what is out there, what are you trying to do? Who are we all trying to kid?

He went on to say that experimental drugs were being discontinued throughout the industry because companies feared lack of reimbursement upon approval. These actions have had an impact on the industry's productivity in the past decade.

Compounds already on the market are not immune to these pressures. To most working in pharmaceutical R&D, it became clear a decade ago that the era of novel antihypertensive research was coming to a close. The blood pressure market was saturated with many safe and effective drugs, most of which were generic or soon to be so. These treatments included diuretics, beta blockers, alpha blockers, angiotensin converting enzyme (ACE) inhibitors, and angiotensin receptor blockers (ARBs). Thus, it was pretty surprising that Novartis pursued renin inhibition, yet another type of mechanism for blood pressure lowering. This approach was not totally novel. In fact, scientists had been working on trying to discover and develop novel renin inhibitors since the early 1980s. For a variety of reasons, most companies had abandoned their efforts by 2000 because it was no longer clear that a renin inhibitor would have any benefit over existing therapies.

Thus, it was pretty surprising that Novartis continued in this field and brought aliskerin (sold as Tekturna or Rasilez) to market in 2007. The fact that Novartis scientists were able to find such an agent after decades of research in an area where many had failed was remarkable. Aliskerin was indeed a potent blood-pressure-lowering drug. It significantly lowered blood pressure for a full 24 hours when given as a single dose, and it added efficacy when dosed on top of other blood pressure medications. But its blood-pressure-lowering effects weren't dramatically better than existing therapies. Was aliskerin too late to market for it to be a commercial success? Renin is an enzyme that initiates the first step in what is called the "renin-angiotensin (RAS) cascade," which ultimately produces the blood-pressure-regulating peptide angiotensin 2. But both ACE inhibitors and ARBs also target the RAS cascade, so many questioned whether aliskerin would offer meaningful advantages over existing agents. Why would payers be willing to pay a premium price for a new unproven agent without superiority data?

Novartis tried to show the medical importance of aliskerin by conducting a number of long-term outcomes studies to demonstrate its advantages. One of these was called ALTITUDE (ALiskerin Trial In Type 2 Diabetics nEphropathy). Novartis described the trial as follows:

> The placebo-controlled Phase 3 ALTITUDE study is the first trial to investigate Rasilez/ Tekturna for more than one year in a specific population of patients with type 2 diabetes and renal impairment. These patients are known to be at high risk of cardiovascular and renal events. In the study, Rasilez/Tekturna was given in addition to optimal cardiovascular treatment including an angiotensin converting enzyme (ACE) inhibitor or angiotensin receptor blocker (ARB).

ALTITUDE was an events-driven study involving 8600 patients, and it was monitored by an independent Data Safety Monitoring Board (DSMB). (A DSMB is a group of independent experts that reviews the ongoing conduct of a clinical trial to ensure patient safety and the scientific merit of the trial.) Novartis was hoping to show that aliskerin, when added to conventional therapy, delayed heart and kidney complications in the type 2 diabetes population. Basically, Novartis hoped that in addition to lowering blood pressure, aliskerin would also have protective effects for organs like the kidney. In studies like this, it is the role of the DSMB to monitor the progress of patients on a periodic basis in order to determine how well the novel treatment is working.

On December 20, 2011, Novartis announced that the DSMB recommended the ALTITUDE study be halted. To great surprise, they found that the trial arm that contained aliskerin after 18–24 months resulted in an *increased* incidence of nonfatal stokes, renal complications, hyperkalemia, and hypotension in this high-risk study population. As a result, Novartis immediately halted promotion of aliskerin-based products for use in combination with ACE inhibitors or ARBs. For some unknown reason, aliskerin doesn't seem to have organ-protective properties, but its use appears to cause unforeseen toxicities when given in combination with other blood-pressure-lowering medications.

This is a terrible result for Novartis and is likely to cause the demise of this drug. Matthew Herper's *Forbes* story[24] on this event quoted Texas cardiologist Dr. John Osborne.

> The Novartis hypertension franchise is now "dead on arrival," obviously. Furthermore, this class of DRIs has died with the death of this drug. . . . Furthermore, given this data, why would one use this molecule anyway?

A few lessons can be drawn from this.

1. Going after a new mechanism for a disease or condition where there are already excellent treatments on the market is always very risky, especially when generics are already present or on the horizon. This isn't limited to antihypertensives. For example, an area like LDL-cholesterol lowering is very

well-served with statins. Any new LDL-lowering agent would have to be very special for payers and physicians to accept it. The same thing can be said for antihistamines, antiulcer agents, and so on. R&D resources are best spent in areas of major medical need, where a new medicine can make a difference.

2. If you still believe that your new drug has something to offer patients, despite the fact that good medicines already exist in this therapeutic area, it behooves you to do key clinical studies before filing the NDA. My guess is that, despite the fact that Novartis has generated income over the past 4 years with aliskerin, the R&D, manufacturing, and launch costs for this drug were pretty high and have not been equaled by the sales to date.

3. This type of result further supports conservatism by regulatory authorities. It wouldn't be surprising if the FDA uses this case as an example of why it wants more Phase 3 studies carried out BEFORE approval to justify registration of a new drug in a therapeutic area already well-served.

When a drug fails, it is not an event that impacts only one company. Presumably, other R&D organizations will learn from this and design future clinical development programs appropriately. For regulatory agencies, one would hope that this type of result won't cause further conservatism and cause them to increase the demands placed upon companies engaged in the discovery and development of new medicines.

One might assume that only the big biopharmaceutical companies are impacted by seeking differentiation of their experimental drugs before NDA filing. Actually, this is a strategy being utilized by even small companies. AVEO Pharmaceuticals is developing a new compound, tivozanib, to treat kidney cancer. AVEO took a gamble and carried out a head-to-head study against Onyx's Nexavar, a leading agent in this field. Tivozanib proved to be superior to Nexavar in this trial in terms of keeping the progression of advanced kidney cancer (Progression Free Survival, or PFS) at bay for longer periods of time. AVEO CEO, Tuan Ha-Ngoc, said of this study:[25]

> We are very pleased by these results, especially the PFS benefit demonstrated in the treatment naïve population, which represents the most significant market opportunity for tivozanib.

What he didn't need to say was that this study will go a long way toward convincing regulators and payers to approve the drug and also get favorable pricing for tivozanib—an important consideration in the already competitive marketplace for kidney cancer drugs.

Sometimes studies designed to differentiate your drug from others can fail, or provide results that question any perceived advantage that your compound may possess. Take, for example, the situation that AstraZeneca (AZ) faces with its successful cholesterol-lowering drug, Crestor.

Rightfully, much has been made of the expiration of Pfizer's US patent for Lipitor (generic name, atorvastatin), which occurred in most countries during 2011. Lipitor is the most notable of the statin class of lipid-lowering drugs, compounds that have helped to lower the rates of heart-disease-related deaths around the world.

Lipitor's commercial success was unprecedented. At its peak, it had worldwide sales in excess of $13 billion annually. Even with its patent expirations in recent years in various parts of the world, it still rang up $9.6 billion in 2011, keeping it the world's number one selling drug.

Lipitor's success can be attributed to a number of factors, but most notably the large (and expensive!) clinical trials Pfizer carried out over a number of years that showed how the treatment of patients at risk of heart attacks and strokes have greatly reduced incidents of adverse events when treated with Lipitor when compared to placebo. Its efficacy, coupled with its remarkable long-term safety profile, has made Lipitor a household name.

When the US patent expired, generic drug makers began selling atorvastatin (the active ingredient in Lipitor) at greatly reduced prices. It is not unusual for the costs of generic drugs to be 10–20% of what the branded drug costs. Thus, Pfizer's sales of Lipitor will drop dramatically, as pharmaceutical industry analysts have noted for the past few years.

What does this have to do with AZ? Well, AZ sells its own statin, Crestor (generic name, rosuvastatin), and with annual sales in excess of $5 billion, it is a major product for the company. AZ has always touted Crestor as the superior compound with slightly better lowering of LDL ("bad cholesterol") and slightly better elevation of HDL ("good cholesterol"). Unfortunately for AZ, Lipitor had been on the market for a number of years before Crestor made it, and the differences in cholesterol modulation between the two drugs was not significant enough for many physicians, who were already very familiar with Lipitor, to switch their patients to the newer drug. Thus, Crestor, while still an important and heavily prescribed drug, trailed Lipitor.

But with generic Lipitor available, AZ has a major problem—one that is a reflection of the cost-consciousness of health-care systems. The generic price of atorvastatin will likely be one-tenth of the cost of Crestor. Thus, payers—insurance companies, Medicare, and so on—will insist that patients newly diagnosed as needing a cholesterol-lowering agent be prescribed the less expensive atorvastatin rather than Crestor. Even more concerning will be those health-care plans that will try to have their Crestor patients switched to atorvastatin in an attempt to reduce their costs. Therefore, the arrival of generic atorvastatin, therefore, not only impacts Pfizer, but also might cause a drop in Crestor sales as well.

AZ has been acutely aware of this issue for years and has tried to take steps to differentiate its drug from atorvastatin. Interestingly, Crestor has a biologic property that differentiates it from atorvastatin: It lowers an inflammatory marker called C-reactive protein (CRP). CRP is associated with a variety of inflammatory conditions, including inflammation of the arteries. Scientifically, it is thought that high levels of cholesterol damage arterial walls and the process of repairing the artery wall results in the beginning of atherosclerosis. High CRP levels could therefore be a signal of early atherosclerosis.

For unknown reasons, Crestor lowers CRP levels, and this distinguishes it from other statins. Is this medically significant? AZ thought so. In an effort to prove it, the company ran the SATURN trial. This study compared Crestor directly with Lipitor in measuring the build-up of plaque in the arteries of patients with heart disease. AZ's thinking, and hope, was that the differences in the biochemical profile of Crestor versus Lipitor would translate into a meaningful clinical difference in slowing the progression of atherosclerosis. If the theory held up, it would be a great result because AZ would be able to clearly show that Crestor was the superior agent.

Early data from the SATURN trial indicated that, while Crestor showed encouraging trends in reducing the fatty deposits in arteries, the results were not statistically significant from the results seen with Lipitor. AZ is carrying out other studies designed to distinguish Crestor. But for now, the claim of Crestor's superiority will likely be nullified. In the face of the dramatic drop in price for generic atorvastatin, it will be hard for physicians to justify prescribing the more costly drug. The Lipitor patent expiration won't impact only Pfizer's net sales in 2012; AZ's Crestor sales are also going to feel the impact.

One would think that, given the high expectations that payers have for new medications, companies would be circumspect about bringing forward new compounds that are intended for an indication already being well-served. This still doesn't seem to be the case. Take avanafil, a molecule being developed by Vivus which is in phase 3 studies for erectile dysfunction. Avanafil is a PDE-5 inhibitor and acts in the same way as do the now famous drugs, Viagra, Cialis, and Levitra. Avanafil is said to be a "new generation" of erectile dysfunction drugs in that it is claimed to have an onset of action of 15 minutes (compared to 30 minutes for the marketed drugs) and is more selective and so should be better tolerated. But, are these differences enough to discourage a switch from the agents already appreciated by patients and their doctors? Furthermore, over the next few years, the patent life for each of the marketed drugs will expire. Given the price drop that will occur, how likely is it that payers will be willing to reimburse patients for premium-priced avanafil when much cheaper (and more famous) generics are available?

CONCLUSION

In the 1990s, when a compound entered phase 3 studies, it generally had a 90% chance of success in terms of completing these late-stage studies and getting FDA approval. But that was a completely different time. In those days, phase 3 served to confirm the results seen in phase 2 proof-of-concept studies. In addition, it often didn't matter if your entry was the fourth or fifth in class. If these disease areas were big enough, all five entries would have some degree of success. Perhaps the top two agents in the class would be billion dollar blockbusters, but the also-rans would be able to garner sales in the $300–$400 million dollar range, enough to justify the investment in the program.

In the "old days" the FDA would often approve a drug based on its impact on disease markers. If your drug lowered cholesterol effectively, it would be approved. The same was true for lowering blood sugar. If the FDA wanted to know about the overall benefits (or risks) of your drug, these would be done after the drug was launched as part of phase 4 studies. Given the results with Avandia and other drugs that have metabolic effects, this is no longer true. For drugs that alleviated pain, you only had to show efficacy and then "long-term" studies of 90 days. The thought of

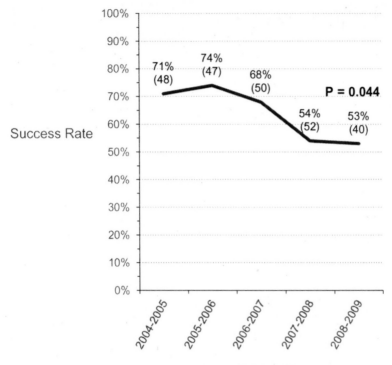

Figure 2.5   New development project success rates. Percentages represent success rate percentage, and values in parentheses represent the number of new development projects assessed in each new range.

running a year-long study looking at the overall physiological effects on osteoarthritis patients was unheard of. Since Vioxx, this, too, is no longer true.

The impact of these higher hurdles being set by regulators and payers can be seen in Figure 2.5. These data, from John Arrowsmith at Thomson Reuters, are taken from a consistent cohort of 18 companies. The dramatic drop in compounds surviving phase 3 is stunning. There are a variety of reasons for this drop-off from the historical 90% success rates to less than 60%. The need for more comprehensive safety and outcomes data is clearly taking a toll on R&D success.

Perhaps some long for the "good old days," but on balance I think we are all better off with the current system as it has evolved. For NDA approvals, the 1990s can be viewed as an era where there were many large organizations producing multiple compounds that didn't need to be differentiated from existing therapy. Nor did these compounds need to show multiple year safety and disease reduction in patients. The rules, however, have changed. The hurdles and costs for new drug development are higher than ever before. Yet, the drugs that do successfully emerge from such a rigorous process are likely to be major advances. Things are definitely more difficult now, but the end result is that patients, physicians, and payers are all better off under the new rules. In addition, biopharmaceutical companies have a great incentive to go after "high value" diseases such as Alzheimer's.

REFERENCES

1. Miller, G. (2010) Is Pharma running out of brainy ideas? *Science*, **329**, 502–504.
2. Paul, S. M., Mytelka, D. S., Dunwiddie, C. T., Persinger, C. C., Munos, B. H., Linborg, S. R., and Schacht, A. L. (2010) How to improve R&D productivity: The pharmaceutical industry's grand challenge. *Nature Reviews Drug Discovery*, **9**, 203–214.
3. Schmid, E. F., Smith, D. A. (2005) Is declining innovation in the pharmaceutical industry a myth? *Drug Discovery Today*, **10**, 1031–1039.
4. Loftus, P. (2010) Merck to Shutter 16 Facilities. *Wall Street Journal*, July 9.
5. Herper, M. (2011) Peering Into Pfizer's Future. *Forbes*, April 10.
6. Pierson, R. (2011) Pfizer trims 2012 sales view, cuts R&D spending. Reuters, February 1.
7. Waters, R. (2010) Shrinking U.S. biotech sector lost 25% of companies in past 3 years. Bloomberg, October 5.
8. Armstrong, R. (2009) Biotech will save Big Pharma. Dow Jones Newswire, May 9.
9. Data generated by Dr. John Arrowsmith at Thomson Reuters indicate that industry average for 2005–2009 for Phase 2 dwell times is 2.8 years (private communication).
10. Arnst, C. (2009) Drug mergers: Killers for research. *Business Week*, March 9.
11. LaMattina, J. L. (2011) The impact of mergers on pharmaceutical R&D. *Nature Reviews Drug Discovery*, **10**, 1–2.
12. Firestone, R. A. (2011) Lessons from 54 years of research. *Nature Reviews Drug Discovery*, **10**, 93.
13. Howard, L. (2010) These days, finding a new job means finding a new career. *The Day*, June 13.
14. Reid, K., Berkrot, B. (2010) Glaxo sees more industry consolidation. Reuters, November 10.
15. Rockoff, J. D. (2011) Pfizer, Merck take different R&D tacks. *Wall Street Journal*, February 4.
16. Stovall, S. (2011) R&D cuts curb brain–drug pipeline. *Wall Street Journal*, March 28.
17. Carpenter, D. (2010) Reputation and Power: Organizational Image and Pharmaceutical Regulation at the FDA. Princeton University Press, Princeton, NJ.
18. Hamburg, M. A. (2011) America's innovation agency: The FDA. *Wall Street Journal*, August 1.
19. Yukhananov, A. (2012) Drug lobby wants clearer FDA rules for diet pills. *Reuters*, January 20.
20. Calabresi, M. (2010) After Avandia, does the FDA have a drug problem? *Time*, August 12.
21. Bolden, M., Swamidass, S. J. (2011) A new bargain for drug approvals. *Wall Street Journal*, July 25.
22. Shimamura, T., Shiroshi, M., Weyland, S., Tsujimoto, H., Winter, G., Katritch, V., Abagyan, R., Cherezov, V., Liu, W., Han, G. W., Kobayashi, T., Stevens, R. C., Iwata, S. (2011) Structure of the human histamine H1 receptor complex with doxepin. *Nature*, **475**, 65–70.
23. Krauskopf, L (2011) Pharma asks the money question earlier for new drugs. Reuters, December 20.
24. Herper, M. (2011) ALTITUDE Study of Aliskiren terminated early by Novartis. *Forbes*, December 20.
25. AVEO Press Release (2012) AVEO and Astellas announce tivozanib successfully demonstrated progression-free survival over sorafenib in patients with advanced renal cell cancer in phase 3 TIVO-1 trial. January 3.

CHAPTER 3

KEY THERAPEUTIC AREAS FOR IMPROVING HEALTH

> Over the past two decades the pharmaceutical industry has moved very far from its original high purpose of discovering and producing useful new drugs. [It is] now primarily a marketing machine to sell drugs of dubious benefits. . . .
>
> —Marcia Angell, *The Truth About Drug Companies*

In July of 2011, I gave a lecture at Drew University. The setting for the talk was a course given there every year at their Residential School of Medicinal Chemistry. This is a one-week session designed to teach new scientists in the biopharmaceutical industry the nuances of drug discovery and development. My presentation was about the misconceptions that exist around the role and contributions to medicine made by the pharmaceutical industry. During the Q&A session, a member of the audience asserted that academic institutions were responsible for roughly 50% of the new medicines that are approved annually. This is a provocative assertion—and completely false.

The topic of who is actually responsible for new medicines is not a new one. Worried that American taxpayers should be sharing more in the profits of the pharmaceutical industry, Congress commissioned a study in 2001 to determine which top-selling drugs had their origins in work done by the National Institutes of Health (NIH). The Department of Health and Human Services prepared the final report which showed that, of the 46 drugs that had annual sales of $500 million or more, only three were associated with Federal patent ties. The other 43 drugs were discovered and developed by the pharmaceutical industry with no Federal investment.

This study is a decade old. Critics of the industry are again challenging the premise that industry is the major driver for the discovery and development of new drugs. An article in *Nature Biotechnology* entitled "Debate Re-ignites Contribution of Public Research to Drug Development" provides some valuable data on this topic.[1] Perhaps the most relevant information is taken from a paper that emanated from Boston University and was published in the *New England Journal of Medicine*.[2] This work, which focused on "The Role of Public-Sector Research in the

Devalued and Distrusted: Can the Pharmaceutical Industry Restore Its Broken Image?
First Edition. John L. LaMattina.
© 2013 John Wiley & Sons, Inc. Published 2013 by John Wiley & Sons, Inc.

Discovery of Drugs and Vaccines," shows that of the 1541 drugs approved by the FDA from 1990 to 2009, nearly 10% were rooted in public sector research. Clearly, the public sector is making important contributions in this field. Yet, the fact remains that the private sector is still responsible for 90% of new medicines.

The NIH, as well as other academic institutions and research institutes, plays a critical role in funding important biomedical research that provides broad benefits not just to the pharmaceutical industry, but to society in general. Furthermore, the fact is that the vast majority of basic biological research is done in academia. But one must distinguish between important theoretical work and the application of this work in discovering and developing new medicines. Basic research is not drug discovery. The NIH does a great job in providing basic knowledge and hypotheses about the nature of living systems. However, it is the pharmaceutical industry, both large companies and small biotech firms, that discovers and tests the compounds to prove or disprove these medical hypotheses. Neither can work without the other. A successful academic–industry partnership is crucial in discovering new medicines.

A recent example in the discovery and development of a drug to treat a form of cystic fibrosis (CF) serves to illustrate the importance of this partnership. CF is the most common lethal genetic disease in Caucasians. Those who inherit CF have lungs with airways that clog with thick mucus, thereby creating a breeding ground for bacteria and making breathing progressively difficult. In 1989, Dr. Francis Collins, the current head of the NIH, and his collaborators found the genetic basis of this disease. CF is caused by defects in a gene that encodes for a protein that is called CFTR. CFTR regulates chloride and water transport in the body and, when it is defective, a breakdown in this transport pathway occurs leading to mucus build-up.

On January 31, 2012, the FDA approved a new drug from Vertex known as Kalydeco (generic name: ivacaftor), which is able to improve the function of CFTR in about 5% of patients who have this disease. It took a long time to get to this point as was explained by Pamela Davis in a *New England Journal of Medicine* editorial entitled "Therapy for Cystic Fibrosis—The End of the Beginning."[3]

> This report is the destination of a long journey that began with the discovery of the gene for cystic fibrosis in 1989 and that has taken us through the definition of the basic defect, the identification of drug candidates by high-throughput screening, the testing in cell, and animal models, and initial human trials to the present gratifying results. . . . This success of ivacaftor is a triumph resulting from the discovery of the cystic fibrosis gene followed by insightful and collaborative basic-science studies conducted by academic and industry investigators that led to clinical trials in an established clinical-research network to produce and validate a novel therapeutic agent for a dread disease.

Without the early basic work done at the NIH, the scientists at Vertex wouldn't have had a starting point. Without the Vertex scientists doing medicinal chemistry to discover a drug candidate, followed by all of the preclinical and clinical studies required to prove that the drug is safe and effective, the CF breakthrough that fulfilled the promise of the early genetic breakthrough never could have been realized.

This is but one of hundreds of examples that can be given of the value that the biopharmaceutical industry adds to improving the world's health. The industry has the resources and talent to discover new medicines across many diseases. Where should these resources be invested? Particularly in the case of the large companies, what therapeutic areas offer the best cases for significant return on the R&D investments that can be made? Equally important, what are the challenges and hurdles that exist in these areas? This part of the book seeks to answer these questions. It is not an attempt to be all-inclusive. The areas that will be focused on are those diseases that plague large segments of the population. But the issues and challenges in each of these therapeutic areas can be considered to be representative of those faced in seeking treatments for uncommon diseases as well.

CANCER

Receiving a diagnosis of cancer is frightening for anyone. Envisioning the medical procedures that are likely to ensue, such as surgery and/or chemotherapy, with no guarantee that these will stave off death, is difficult to endure. Yet, treatment options are far better now than 25 years ago when the primary options for chemotherapy involved dosing with cytotoxic agents. These chemotherapeutants had been designed to kill fast-dividing cancer cells. Unfortunately, these compounds can also damage normal cells, and this relative lack of cell selectivity is what causes the severe side effects associated with these agents.

However, new agents are now available that offer promise to cancer patients. These new medicines have their roots in decisions taken 40 years ago. In his 1971 State of the Union address, President Richard Nixon described his "War on Cancer," asking the nation for levels of effort similar to those used on the Manhattan Project or the Apollo Lunar Program. On December 23, 1971, he signed into law the National Cancer Act, which committed an additional $100 million to the National Cancer Institute for cancer research.

The basic research funded by the "War on Cancer" increased the understanding of cancer molecular biology such that mechanisms specific to tumor growth were elucidated. Research showed that cancer was not a single disease but rather a broad set of diseases resulting from aberrant cell growth caused by mutations in a variety of biological pathways. The triggers for this aberrant cell growth are known as "oncogenes," genes that, when activated, turn normal cells into tumor cells. This research, which evolved over the next 15–20 years, provided drug hunters just what they needed, namely, potential targets for novel drugs that, if effective, could specifically destroy tumor cells while causing little, if any, damage to normal cells. In essence, these potential drugs could be more effective in treating cancer than the older cytotoxics and could prove to be safer as well.

Companies both large and small began to invest heavily into cancer research in the 1990s, resulting in an explosion of potential new medicines for cancer. The pipeline for new cancer drugs has grown across the biopharmaceutical industry unlike anything ever seen. The Pharmaceutical Research and Manufacturing Association estimates that there are about 1000 novel anticancer agents currently in clinical development. Not all of the experimental medicines will make it through the development process, receive regulatory approval, and become drugs. But the odds are that 100–200 or more will—a tremendous accomplishment for the biopharmaceutical industry after long years of research. In fact, it seems that a new breakthrough is reported in the press on a weekly basis involving new compounds to treat kidney cancer, pancreatic cancer, melanoma, breast cancer, lung cancer, and so on. These agents also act in a variety of different ways. Some block a specific step in the biological cascade that allows the tumor to proliferate. Some agents starve a patient's tumor by preventing the tumor from growing blood vessels (a process known as anti-angiogenesis), thereby preventing it from getting nutrients. Other compounds are designed to reawaken one's immune system to help fight the cancer.

As described by Harpal Kumar, the chief executive of Cancer Research UK, we are in a "golden era" of research.

> There has been an explosion in our understanding of what cancer is, why it happens, why it doesn't happen in some people, and why it moves around the body.[4]

The progress being made and the changes in how novel cancer drugs are being developed can best be exemplified by the story of crizotinib (sold by Pfizer as Xalkori), a drug to treat a subset of patients with lung cancer. It had mistakenly been assumed that lung cancer was a single disease. But, while under a microscope all lung cancers might look the same, the basis for their formation can be very different. Crizotinib is an inhibitor of anaplastic lymphoma kinase (ALK). There are some patients diagnosed with lung cancer who have never been smokers. Genetic studies have shown that their lung cancer is a result of the ALK gene, working in concert with another gene, staying on constantly in lung cells, thereby fueling uncontrolled growth. Crizotinib is able to stop that growth. Even better is that there is a diagnostic test that can determine if a patient's lung cancer is driven by the ALK oncogene. As a result, only those patients who will benefit from crizotinib will be exposed to the drug, thereby saving both the patients' and doctors' time and efforts and the healthcare system significant costs. This is important because less than 10% of lung cancer patients have tumors that are susceptible to this drug.

This represents a new paradigm in the way drugs can be developed. Crizotinib was first nominated for development in 2005. It was on the market in 2011. The ability to target a specific set of patients, as well as show that a drug is safe and effective, can occur relatively rapidly and at less expense than usual with a targeted drug like this.

There are other types of cancers that appear to be driven by the ALK oncogene, such as ALK-positive non-Hodgkin's lymphoma and inflammatory myofibroblastic tumor (IMT).[5] Studying crizotinib in these cancers should be straightforward in that these patients will first be tested to see if their cancer is being driven by this oncogene. Thus, in this type of specifically targeted drug trial, fewer patients should be needed to show that crizotinib is indeed effective in eliminating their disease.

The crizotinib story is not unique. There are other such targeted therapies now on the market, with more on the way. Furthermore, as was alluded to earlier, these drugs act by mechanisms that should be able to be used in combinations that complement each other. In fact, many researchers now believe that, much like the case with

AIDS, we are not far from the time when a cancer diagnosis will not be a death sentence but rather a diagnosis of a chronic disease, one that might not necessarily be cured but rather be treated with a variety of medications that will keep one's cancer in check. As AIDS patients are treated with a combination of drugs, so too, will cancer patients be treated. Getting to this point will be the result of decades of research and billions of dollars in investment. But what a wonderful situation for patients.

The new cancer medications are expensive, and annual costs can range anywhere from $10,000 to $100,000. Furthermore, multiple drugs will be needed to keep one's cancer in check. As was eluded to earlier, one can envision a patient being on (a) a drug specifically designed to stop cancer cells from proliferating in her breast and (b) another drug designed to prevent the breast tumor from growing new blood vessels, thereby starving it, and (c) a third drug designed to help her own immune system to fight the disease. There are now drugs on the market that can do all of this.

This welcome news for cancer patients, unfortunately, poses a potential problem for health-care systems. Conversion of cancer from a fatal disease to a chronic one means that many people (millions?) will be living for 10, 20, or 30 years on expensive medications. One can begin to do the math and show the tremendous burden that such costs will add to an already high health-care bill for governments and other payers. One can argue that, as the cancer survivor population grows, paying for its medicines will become unsustainable. This is being referred to as "financial toxicity". How does one get to a situation where all patients will be able to get lifesaving medications?

The companies that will be successful in this field will be ones that can "bundle" the medications needed to treat certain cancers. Thus, it will behoove companies to develop, either internally or through alliances, the best drugs to treat certain cancers. By having these multiple agents in hand, one can envision a scenario where a company can work with health-care providers to deliver triple therapies (drug "cocktails") at a reasonable cost. There is evidence that such positioning is already happening. BMS and Roche have joined forces to combat melanoma. They have agreed to fund clinical studies that combine BMS's recently approved melanoma drug, ipilimumab, with Roche's compound, vemurafenib. Both of these attack melanoma in different but potentially complementary ways. Theoretically, the combined therapy should be far more effective than either is alone. Perhaps it is naïve to assume that, should this combined therapy be successful, the cost of this combination would not be additive but rather be lower than the cost of each individually. Nevertheless, one would expect more such joint ventures in the future; and as multiple cancer treatments emerge, perhaps competition will help drive down price.

The "War on Cancer" has been waged for decades. Now that it appears that the control of this disease is possible, it is imperative to find business models that allow for patients to be treated with medications and protocols that don't bankrupt the health-care system.

A second issue that can stymie progress in developing new cancer treatments is the reduction of R&D investment in this field. The cutbacks in biopharmaceutical R&D, as outlined in Chapter 2, will have an impact. Equally concerning, however, are the budget cuts being made at the NIH as Congress tries to trim the US budget deficit. These concerns were voiced by Dr. Judy Garber, Director of the Center for Cancer Genetics and Prevention at the Dana Farber Cancer Institute.[6]

> We finally invested enough in the infrastructure to see an accelerating pace of break-throughs . . . [such targeted drugs] are the tip of the iceberg. After we've finally gotten to this point, do we say, "We've got these two, that's enough?" We'd be foolish to miss the opportunity we now have.

This is a valid concern. Tremendous progress has been made in the treatment of cancer over the last 40 years. Data from the American Cancer Society shows that, for all cancers combined, the mortality rates (per 100,000) for men were 280 in 1990, and those for women were 175. By 2006, this dropped to 221 and 154, respectively.[7] A good deal of this improvement can be attributed to the decline in smoking and early detection methods. But the new medicines that are emerging from the biopharmaceutical industry are making a big difference. Nixon's vision has yet to be realized. But we are on the cusp of being able to convert cancer to a treatable disease.

DISEASES OF THE BRAIN

A while back, the US Congress designated that the 1990s would be the "Decade of the Brain." Their intent wasn't unreasonable. After all, diseases like depression, schizophrenia, and Alzheimer's disease (AD) are major problems for society. But despite Congress's good intentions, not a lot of tangible progress was made in the 1990s because these are very difficult diseases to conquer. Perhaps it would have been better to christen the twenty-first century as the "Century of the Brain," because it is going to take more than a decade to bring new medicines to patients suffering from these ailments. The brain is proving to be the last great frontier of biology.

One can broadly characterize diseases of the brain into two categories: psychiatric diseases such as depression, schizophrenia, and anxiety; and neurodegenerative diseases, the major one being AD. Scientific pursuit of each poses distinct issues. Take, for example, clinical trials designed to determine whether experimental medicines are efficacious. For psychiatric diseases, studies generally run for 8–12 weeks in order to determine if a drug works. In addition, activity in rodent models can be predictive for human efficacy. For AD studies, where one hopes to either halt or reverse the disease, the studies may need to run for 12–24 months. Furthermore, it is not yet clear that activity in animal models is at all predictive for treating the human condition.

The use of psychiatric medications continues to be controversial, in terms of both their need and their effectiveness. As outlined in Chapter 1, antidepression drugs are a major target of these criticisms. Yet, despite this, as reported in the *Wall Street Journal*,[8] use of drugs to treat mental illness has grown to such an extent that one in five adults is now taking at least one psychiatric drug. The primary growth is coming from increased use of antipsychotics and attention deficit disorder medicines. This is not just a US problem. BBC News reported in 2011 that one in 10

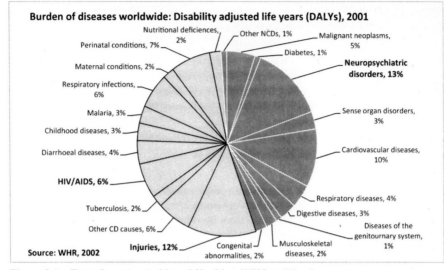

Burden of diseases worldwide: Disability adjusted life years (DALYs), 2001

Nutritional deficiencies, 2%
Perinatal conditions, 7%
Maternal conditions, 2%
Respiratory infections, 6%
Malaria, 3%
Childhood diseases, 3%
Diarrhoeal diseases, 4%
HIV/AIDS, 6%
Tuberculosis, 2%
Other CD causes, 6%
Injuries, 12%
Source: WHR, 2002
Congenital abnormalities, 2%
Musculoskeletal diseases, 2%
Diseases of the genitournary system, 1%
Digestive diseases, 3%
Respiratory diseases, 4%
Cardiovascular diseases, 10%
Sense organ disorders, 3%
Neuropsychiatric disorders, 13%
Diabetes, 1%
Malignant neoplasms, 5%
Other NCDs, 1%

Figure 3.1   From *Investing in Mental Health*, a WHO publication.

Scots are on antidepressants. Blogger Paul Southworth ("Humanospere"), a visiting scholar at the NIH, calls mental illness "the most neglected disease in all of global health." He makes his point by utilizing "disability adjusted life years" (DALYs), which measures the impact of diseases, injuries, and disorders on survival. Using data from the World Health Organization, he showed that mental illness is responsible for 13% of global DALYs—more than cardiovascular diseases (10%) and cancer (5%) (Figure 3.1).

Dr. John Mann, in his *New England Journal of Medicine* article[9] entitled "The Medical Management of Depression," presents data that support Southworth. Major depressive disorder, a much maligned disease, accounts for 4.4% of the total overall global disease burden. Further sobering are the following facts. Patients who have diabetes, epilepsy, or ischemic heart disease with concomitant depression have poorer outcomes than those without depression. More concerning is that the risk of death from suicide, accidents, heart disease, respiratory disorders, and stroke is higher among the depressed.

There are a variety of medications available to treat depression and other psychiatric illnesses. But these are not ideal. In general, the success rate for these drugs is on the order of 50%. In addition, many take weeks of treatment before the beneficial effects begin to take effect, thus prolonging a patient's medical problems. Finally, these drugs have a variety of side effects that impact one's quality of life.

Being that this is still an area of major medical need and where there are large patient populations, one would think that drug companies have large teams of scientists working on the next generation of novel psychiatric medicines. In fact, the opposite is true. Companies like Novartis, GSK, and AstraZeneca have closed their neuroscience research efforts. Other large pharmaceutical companies such as Merck, Pfizer, and Sanofi have drastically scaled back their efforts as well. There are a variety of reasons for this. For one thing, clinical trials in this field are often risky due to the high placebo effects seen in these studies. It is very disheartening to run an expensive clinical program, only to find that the high placebo success rate resulted in little difference between it and your drug. Often companies need to run multiple studies in order to find the necessary two successful trials needed for FDA approval. Furthermore, given the fact that most psychiatric drugs are now generic, the hurdle for a new drug is even higher as payers aren't interested in a new drug unless it is superior to the existing generics.

Needless to say, there are many who are concerned about this pullback in research. Mike Williams in *BioWorld* has this to say:

> As is typical in situations in biopharma where considerable investment has resulted in failure, the response has been for the pendulum to swing from one extreme to another. In this instance, the extremes range from the decidedly Pollyanna-ish views emergent from the Decade of the Brain to the view that the translational "valley of death" for CNS is actually a "cesspool of devastation," reflecting a grievous disconnect between the promise, accomplishments, discoveries and billions of dollars spent in contemporary neuroscience . . . [and] . . . the abysmal lack of conceptually novel therapeutics in psychiatry.

Williams isn't alone in his views. Professors David Nutt (Imperial College London) and Guy Goodwin (Oxford University) have recently authored a report for the European College of Neuropsychopharmacology which sounds the alarm on the elimination of research by pharmaceutical companies in areas such as depression and schizophrenia. They are greatly concerned that major companies "see research into better neuropsychiatric drug targets as economically nonviable." Their fear is that if such research is stopped, "we will have a dead space of 20 to 30 years before we can begin to do it again."

First of all, it is encouraging to hear prominent academics extol the importance of research that occurs in the pharmaceutical industry. Usually, industrial drug R&D is minimized by people like Dr. Marcia Angell, former editor of the *New England Journal of Medicine*, who promote the view that pharmaceutical companies do little original R&D and instead license all of their drugs from universities and small start-up companies. It is a pleasant change to hear these professors recognize the value of the contributions of the R&D of pharmaceutical companies.

Nutt and Goodwin are justified in their concerns. However, their rationale as to why pharmaceutical companies are abandoning this research is not entirely correct. Because the biology of the brain is so complex, it is not unusual for an agent in the central nervous system (CNS) to have side effects such as changes in mood, anxiety, insomnia or even create suicidal tendencies. Nutt and Goodwin say the industry is spooked by potential litigation over these types of side effects, but every drug causes side effects. If a pharmaceutical company worried exclusively about adverse reactions, it would never develop any new medicine for any condition. As was said earlier, the key in navigating side effects is that everyone involved—the industry, the FDA, the doctors and, most importantly, the patient—understands the risk–benefit profile of a new drug.

Nutt and Goodwin should be applauded for calling attention to the beneficial research done by the pharmaceutical industry; however, the real issue here isn't fear of litigation. It's that there are not many compelling targets for industrial scientists to study. Traditionally, academic researchers have helped to fuel the generation of new ideas by doing basic research into how the brain works. This preliminary research yields areas for industrial scientists to probe. But funding of this type of academic research by institutions like the NIH has decreased over the years. Furthermore, the NIH is now diverting hundreds of millions of dollars to doing drug discovery, which is clearly better done by industry.

To avert the "dead space" feared by Nutt and Goodwin, a greater emphasis on basic CNS research needs to occur. While large companies have gotten out of this area, there are small companies still doing research in seeking new pathways to treating CNS diseases. For example, Sage Therapeutics, a new CNS start-up, has raised $35 million to explore new modulators for key targets in schizophrenia, pain, depression, and brain injury. The FDA needs to express an openness that new medicines that result from this work will be given a high priority in their review process. In addition, the public sector needs to increase funding for academic researchers in the CNS field to plant new seeds that can be developed by industrial scientists.

There is no doubt that despite the medical needs, the topic of psychiatric drugs stirs intense debate. A *New York Times* opinion piece by Dr. Peter Kramer entitled "In Defense of Antidepressants"[10] is a typical example. Kramer, clinical professor of psychiatry at Brown, provided a spirited defense of the current methodology in treatment paradigms. As one would expect, the debate continued a week later with a number of letters to the *New York Times* not just from Dr. Marsha Angell, but also from a number of heads of important psychiatric associations as well as from psychiatry professors from major universities. However, the letters that made the biggest impression on me were the ones from actual patients. One in particular stood out.

> As a professional ethicist, I share . . . concerns about the medical–pharmaceutical complex and how the obsession with ever-greater profits can hinder, not promote, ethically intelligent patient care. But, as someone who has been using antidepressants successfully for many years, I can say from experience that some of that concern is misplaced. My life is richer and infinitely more satisfying because of this medication. I offer my profound gratitude to the dedicated researchers and conscientious clinicians who have made this possible.

These are important drugs. They add value to society. Yes, they need to be prescribed appropriately. But it is important to recognize the need for the next generation of these medicines and support the efforts needed to do so successfully.

Unlike the pullback in the biopharmaceutical industry in R&D for psychiatric drugs, the situation for Alzheimer's disease (AD) is quite the opposite. Most of the industry is expending lots of resources in trying to discover drugs to treat this disease. This is no surprise. The greatest challenge facing the world in terms of health care is AD. As the baby boomer generation ages, the incidence of AD is going to rise dramatically. Today there are 5.4 million Americans with AD. By 2050, that number will triple. The Alzheimer's Association estimates that the cost of care of

AD patients is over $180 billion annually. By 2050, these costs will have devastating effects on the economy unless some medical advancements can be made.

The Obama administration recognizes the threat that AD is posing on the long-term US health-care goals. In a move reminiscent of Nixon's "War on Cancer," the President announced that an additional $156 million would be added to the NIH budget over 2012–2013 to support both basic and clinical research designed to identify genes associated with AD progression as well as to improve therapy. These funds would be added to the existing NIH budget of $450 million annually for AD research.

Interestingly, this may seem like a major commitment. But, according to the Alzheimer's Association, the NIH spends roughly $3 billion on AIDS, $4.3 billion on heart disease, and $5.8 billion on cancer research. Given the growing incidence of AD, one wonders if the NIH should do some reprioritizing in its funding.

If you have friends or relatives with AD, you are undoubtedly hoping that some breakthrough will be made that will alleviate or even reverse this disease in your loved one. Unfortunately, recent reports might dampen your hopes that a breakthrough is right around the corner. The good news is that great progress has been made in understanding the root cause of this disease. We now know that AD is caused by the build-up of proteins into clusters that clog up nerve cells in the brain. These "plaques" break down nerve cells; this results in a decrease in the ability of these cells to function, leading to the familiar AD symptoms of memory loss and an erosion of cognitive skills.

These insights have led researchers to design drugs that prevent the build-up of these plaques. This has not been trivial to do. Over the last 20 years, great work has been done to understand the multiple mechanisms on how these plaques form, develop genetically modified mice that mimic the human condition of AD, and design and synthesize compounds that could block or reverse this process in the human brain. In short, scientists are doing everything they can to try to cure the disease. A number of companies like Lilly, BMS, J&J, and Pfizer have compounds in late-stage development in AD patients, and the much-anticipated clinical trial data are starting to become available. But despite the knowledge of AD's cause and the multi-tiered approach to finding a cure, early results have been discouraging.

Interestingly, this is the same place that cancer research found itself in the 1990s. At that time, a great deal of knowledge had been accrued on the genetic basis of tumor formation. The biopharmaceutical industry simultaneously created compounds to test these theories in clinical trials. These trials answered a lot of questions, disproved some theories, and supported others. Eventually, this cutting-edge research led to the new drugs that have been introduced in the last few years. Research in AD seems to now be in that stage.

Lilly has led the field in an area of research known as gamma secretase inhibitors. Lilly hoped that its lead compound, semagacestat, would inhibit the enzyme that is believed to contribute to build-up of the amyloid protein, which, in turn, forms clumps in the brain and causes AD. It was hoped that testing this compound in AD patients would slow or reverse the progress of AD. Unfortunately, Lilly halted the phase 3 trial for this compound because it led to a *worsening* of cognitive function compared to placebo in the AD patients in the study.

Was the Lilly compound flawed in some way? Or is the flaw in the hypothesis that a gamma secretase inhibitor should be of value in treating AD? That answer is not clear. More data will emerge from the study that BMS is doing with its gamma secretase inhibitor, which they feel is superior to semagacestat. Will this compound behave similarly? Perhaps. But this is the nature of trying to discover a drug to treat such a difficult disease.

Another way to attack AD is to try to administer an antibody that can bind to the deposited amyloid and clear it out of the brain. Such an antibody is bapineuzumab, which is being developed by both Pfizer and J&J. Unfortunately, this antibody caused vasogenic edema—brain swelling—in AD patients. Initially, this caused a lot of worry and called this mechanism into question as well. However, subsequent studies are suggesting that such swelling *may* be an indication that the drug is working and that, theoretically, the swelling is a result of blood vessels becoming leaky when ridding the brain of the amyloid protein. Is this a side effect or an indication of efficacy? Obviously, more studies are needed. The FDA is working closely with those running these studies to help find the answers without jeopardizing patient safety. This is yet another example of how progress is made—not necessarily with a major "aha moment," but rather with key experiments pointing you in one direction or away from another.[11]

There are other issues that researchers are trying to navigate. Most of the drugs in clinical trials are in patients with well-established disease. Perhaps their disease has progressed too far for any compound to show cognitive benefits. Thus, maybe trials should be done in patients for whom disease is at the earliest stages. The downside to this practice is that to see a significant effect, the trials will likely need to be of a multi-year duration, which are costly and challenging to do.

With new treatments for AD years in the future, are there steps that people who are at risk of getting AD can take? A *New York Times* article called "Grasping for Any Way to Prevent Alzheimer's" tried to address this.[12] People have touted all sorts of prevention paradigms: weight loss, dietary supplements, and physical activity among them. Unfortunately, an NIH panel recently concluded that none of these has proven to be effective. The issue again stems from the fact that this is a disease that is decades in the making. Studies of any kind to prove that a new treatment or regimen is effective takes years of research.

The biopharmaceutical industry has been working for decades trying to understand AD and develop medicines for it. Great work has been done in understanding the multiple mechanisms on how plaques form and then coming up with molecules that can stop disease progression, if not reverse it. However, the challenge in AD studies is that halting or reversing plaque progression doesn't occur overnight. Thus, clinical trials are long, and it is not until you finish phase 3 that you know if your drug worked.

Before you get too depressed by this, remember that this is the same situation scientists faced in coming up with new treatments for cancer decades ago. However, the explosion in new cancer drugs is rooted in the R&D done back then. The same process is occurring now in AD. Tremendous breakthroughs in science are occurring. It is just going to take a lot more work to cure this disease, and the biopharmaceutical industry is playing a major role in this effort.

CARDIOVASCULAR DISEASE (CVD)

In an editorial focusing on the R&D opportunities and progress being made in cardiology,[13] Peterson and Gaziano make the following observation about atherosclerosis.

> Only 50 years ago, atherosclerosis was thought to be inevitable, a natural consequence of the aging process. However, carefully performed epidemiologic studies from the Framingham Heart Study and others identified the major CVD risk factors including hypertension, elevated cholesterol levels, smoking, and diabetes. These seminal works changed the view of CVD from a preordained fate to a preventable disease.

Much of this progress is the result of the discovery and development of multiple medicines to lower blood pressure and LDL cholesterol. As was discussed earlier in this book, the safety and efficacy of these agents, combined with the fact that many are now available as cheap, generic medications, make it very difficult to bring forward new drugs to treat hypertension and hypercholesterolemia.

This is not to say that all research has stopped in these areas. There is a new target to lowering LDL cholesterol that is generating a great deal of interest called PCSK9 (proprotein convertase subtilisin/kexin type 9).[14] This protein binds to LDL receptors preventing them from recycling in the liver. As a result, LDL levels remain high in the plasma. Interestingly, people born with a genetic defect that prevents them from producing PCSK9 have very low LDL cholesterol levels and little evidence of atherosclerosis.[15]

As a result of the believed importance of this protein in lipid metabolism, a number of companies are looking for compounds that block its actions or formation. The former head of R&D at Amgen, Dr. Roger Permutter, said "It is the biggest opportunity for lowering cardiovascular risk for the entire pharmaceutical industry."

Yet, despite this excitement, there is no guarantee that a drug that will emerge from this field of research will be a statin-like commercial success. For one thing, the experimental medicines furthest in development are antibodies to PCSK9. In all likelihood, they would need to be injected once or twice a month to be effective in patients. Beyond being less convenient to take than statins, these antibodies will likely be very expensive—perhaps costing $10,000 to $20,000 annually. Why would any health plan pay such an exorbitant amount of money in the face of having generic atorvastatin available?

The value of a PCSK9 antibody would be in patients at high CVD risk and for whom statin therapy is either not effective or not tolerated. Thus, while a PCSK9 antibody might be extremely efficacious in lowering LDL cholesterol, it would need to have dramatic advantages over generic statins before being reimbursed by payers. This is yet another example of the impact that payers are having in the discovery and development of new medicines.

While there are good options currently available to treat high LDL, raising HDL, the "good cholesterol," is still a major challenge. In fact, in an editorial in the *Journal of the American Medical Association*, Dr. Christopher Cannon referred to this as the "Holy Grail."[16]

For more than 3 decades, since high levels of high-density lipoprotein cholesterol (HDL-C) were first linked to a lower risk of developing cardiovascular disease, the notion of raising HDL-C levels has been regarded as a potentially ideal treatment to prevent cardiovascular disease. High-density lipoprotein cholesterol has generally been called the "good cholesterol" to distinguish it from low-density lipoprotein cholesterol (LDL-C), which has been clearly linked to increased risk of cardiovascular disease and mortality. Although multiple large randomized trials have shown that lowering LDL-C leads to a reduction in cardiovascular events and mortality, for HDL-C the translation from these observational studies to identifying a drug in randomized clinical trials that both increases HDL-C and reduces clinical events has been long and difficult. As such, the search for an HDL-C-raising, cardioprotective drug almost seems like the quest for the Holy Grail.

The problem that scientists have faced in this field is that not a lot is known about raising HDL, and so trying to accomplish this feat in the laboratory has proven quite challenging. An exception involves the emerging biology around a protein known as cholesteryl-ester transfer protein (CETP). Back in 1990, a report in the *New England Journal of Medicine*[17] described a Japanese family who, due to a genetic defect, didn't produce CETP and as a result had very high HDL levels and low incidence of cardiovascular disease. Because of this observation, a number of companies, including Pfizer, sought to come up with an inhibitor of CETP to test in heart patients. The rationale was pretty simple: By inhibiting CETP, the new drug would raise HDL, thereby mimicking the situation in the Japanese family who were born without this protein. Theoretically, the CETP inhibitor would reduce heart disease maybe as well as statins. In a dream scenario, the combination of statins and CETP inhibitors would reverse heart disease, enhancing the quality of life of hundreds of millions of people.

Finding inhibitors of CETP proved to be very challenging, and most companies gave up. Pfizer almost did the same. But through hard work, creativity, and a bit of luck, the CETP inhibitor torcetrapib was found. It successfully cleared preclinical toxicology studies and entered clinical trials in 1999, eight years after the discovery program began. The initial clinical results were astonishing. Patients on torcetrapib had increases of HDL of 100% or more—an unprecedented finding. Even better, torcetrapib also caused a modest decrease in LDL; however, when torcetrapib was added to Lipitor (atorvastatin), not only was the HDL elevation maintained, but LDL drops of 40–60% were seen. The lipid remodeling that was achieved with this combination had never been seen with any therapy.

The combination of torcetrapib and atorvastatin (T/A) became a major focus not just for Pfizer but for the entire cardiovascular field. As the early clinical data for T/A emerged, other companies that had previously given up their efforts in looking for CETP inhibitors jumped back in the hunt. But although T/A uniquely modified the lipid profile of patients with heart disease, its ultimate value was not yet established. The long-term benefits of altering heart disease with CETP inhibition were still hypothetical. Furthermore, at the end of the phase 2 studies, it was found that T/A caused a small but reproducible increase in blood pressure. However, based on the known association between atherosclerosis, blood pressure, and cardiovascular disease, the large increase in HDL was expected to provide benefits that would be far greater than any harm caused by the small blood pressure increase. To prove this, Pfizer had to carry out an extensive phase 3 program that included a multiple-year study in patients with heart disease with the hope of showing that patients with heart disease fared better on T/A than they did with Lipitor (atorvastatin) alone. This was a pretty high hurdle. Pfizer wasn't content to test T/A versus a placebo; rather, it chose to test T/A against Lipitor, the premier statin that had a successful track record in reducing cardiovascular disease.

The phase 3 program wasn't cheap. Overall, it was projected to cost $800 million. But such an extensive and expensive plan was necessary. Pfizer realized that even if this program was successful, T/A would be competing with generic statins such as simvastatin and eventually atorvastatin itself. For physicians to be willing to prescribe T/A, for payers to be willing to pay for T/A, and for patients to be willing to take T/A, the clinical program needed to show that it was a drug of clear value.

By now, most know of the result of the phase 3 outcomes trial.[18] In late 2006, this study was halted by the Data Safety Monitoring Board (DSMB), which was responsible for overseeing the study. The reason for halting the trial was an imbalance in all-cause mortality in the T/A patients as compared to those on atorvastatin. In other words, T/A was not better than Lipitor alone; in fact, it was worse in reducing heart attacks and stroke.

This result stunned cardiologists around the world. One of the more telling comments was from the renowned Dr. Steven Nissen of the Cleveland Clinic, who said: "These studies further demonstrate the great difficulty in developing therapies to disrupt the atherosclerotic disease process."

As was stated above, it was generally believed that the dramatic HDL-elevating effects of torcetrapib would be of far greater clinical importance than the relatively minor blood-pressure-elevating effects caused by torcetrapib. Furthermore, the patients in this study were already on blood-pressure-lowering medications, thus their hypertension was being controlled. This was not just Pfizer's belief. Some of the world's leading cardiologists, like Dr. Nissen, were not only advising Pfizer on the torcetrapib program, but they were also carrying out the clinical trials. Raising HDL was viewed as the new frontier in cardiovascular research. Heart physicians around the world were anxiously awaiting the results from the torcetrapib studies.

When the torcetrapib trial was halted, everyone in the field was shocked. Suddenly, the promise of CETP inhibition had evaporated. Experts began to question whether raising HDL would have *any* benefit in patients with heart disease. Merck, which had planned to launch their phase 3 program at about the same time that Pfizer halted theirs, instead put their program on hold for a year as they tried to evaluate what to do. Yes, anacetrapib didn't raise blood pressure, but perhaps the CETP mechanism was flawed. After a year of intense internal debate, Merck decided to go ahead with anacetrapib clinical studies, but in a slower fashion than they originally had planned. Their first phase 3 trial, reported at the American Heart Association meeting in 2010, showed that anacetrapib can be safely administered to patients with heart disease for 18 months with dramatic HDL elevation and LDL lowering without undue consequences. They are now running the crucial large-scale phase 3 outcomes study that, hopefully, will show that CETP inhibition can confer additional benefits in heart patients when compared to statin therapy alone.

Will this "Holy Grail" be found? Many people—scientists, patients, physicians—all hope so. But this quest is typical for the biopharmaceutical industry and in some ways mirrors the previous discussion in AD. To prove the value of a CETP inhibitor, long outcome trials are needed, trials that are costly. However, should Merck have success, the search will have been extremely worthwhile. The biopharmaceutical industry is tasked with proving scientific hypotheses, a challenge that more often than not is frustrating. But when successful, not only does this company win but patients do as well.

DIABETES

A recent study in *The Lancet*, a British medical journal, showed that since 1980 the incidence of type 2 diabetes has doubled globally to 347 million and tripled in the United States.[19] The Center for Disease Control (CDC) now estimates that 1 in 12 Americans have this disease. Things are not projected to get better. This International Diabetes Federation predicts that 522 million around the world will suffer from diabetes by 2030. Instead of triggering alarms among those focused on health-care issues, this scientific paper seems to have received only modest attention. Gautam Naik did cover it in the *Wall Street Journal*; and his quote from one of the study's authors, Professor Majid Ezzati, was on the mark: "Diabetes is a long-lasting and disabling condition, and it's going to be the largest cost for many health-care systems."[20]

It is not as if this epidemic has a mysterious cause. The growth in the incidence of type 2 diabetes is directly related to obesity. As was discussed in Part 2, the CDC has been following the girth of America for the past two decades and the results are startling. As people get heavier, the cells in their body are less able to utilize insulin. This inability leads to increased levels of sugar in the bloodstream, which, if left untreated, can result in vascular complications leading to heart disease, kidney failure, and blindness.

The obvious solution to this problem is getting people to exercise more, eat less, and embrace a healthier lifestyle. While admirable efforts to achieve these goals are being made on multiple fronts, they aren't working. Thus, people are going to need to have the option of drug therapy to help alleviate the symptoms and ward off the deleterious downstream effects of diabetes. But this solution isn't so straightforward. While there are drugs currently available to treat diabetes, their effects are modest. Furthermore, two antidiabetic drugs, Avandia and Actos, have been found to be deficient in terms of their risk–benefit profile; as a result, the use of both is extremely limited by regulatory authorities.

Unfortunately, the R&D pipeline of potential new antidiabetic drugs is not overwhelming. While great drugs have been found over the years to control high blood pressure and high cholesterol, diabetes drug discovery has proven to be much more difficult. Here is a telling example.

Bristol-Myers Squibb and AstraZeneca have been co-developing a promising antidiabetic called dapaglifloxin, an inhibitor of the sodium glucose co-transporter (SGLT2). This transporter is responsible for secreting glucose into the bloodstream. Given that diabetics have too much sugar in their blood, a drug that prevents sugar entry in the blood could have long-term benefits of limiting the downstream diabetic complications such as kidney disease, blindness, or amputations.

Inhibition of SGLT2 by dapaglifloxin was thought to be inherently safe since people with a rare genetic disease that results in nonfunctioning SGLT2 are healthy; their only abnormality is very high glucose levels in their urine. An anticipated benefit of dapaglifloxin is that its mechanism of action is independent of insulin. Given that many diabetics have such severe disease that no single agent is sufficient to bring it under control, dapaglifloxin, which lowers blood sugar levels in its own right, would be an ideal drug to add to existing drug regimens.

Unfortunately, an FDA advisory committee, which was convened to evaluate the potential approvability of this drug, voted against recommending it by 7 to 4.[21] Their reasoning was based on the drug's safety profile. About 0.4% of women taking the drug got breast cancer compared to 0.1% in the control group. An excess of bladder cancer was also seen in men (0.3% on drug vs. 0.05% in controls). Is this increase in cancer real? Is it related to SGLT2 inhibition or is it unique to dapaglifloxin? There are other companies that have SGLT2 inhibitors in development. Assuming that they continue moving their development programs forward, answers to these questions will be found. In addition, the FDA may ask developers of compounds in this mechanistic class to carry out multi-year studies to show the benefit of SGLT2 inhibition in the prevention of cardiovascular disease as well as to ascertain if the increase in cancer rates is related to this mechanism.

Given the need for new treatments, new drugs for diabetes must be viewed as a major priority for all involved in the discovery and development of new medicines. As such, R&D in this area needs to be given a much higher priority, particularly by governmental agencies. Here are a few suggestions:

1. More needs to be invested in basic research—The NIH sets the national health priorities where by it invests its budget. Of its $30 billion projected research budget for 2012, the NIH only allotted $1.6 billion to diabetes research. In contrast, the cancer budget is in excess of $5 billion. The investment in cancer research by the NIH over the past three decades has helped to produce a spectacular pipeline of almost 1000 new anticancer agents currently in development, certainly indicative of the priority this area of research has received. But even the infectious diseases budget is four times that of diabetes. Given the enormous prevalence of diabetes, perhaps some redistribution of NIH funds is in order so that more research into understanding diabetes disease mechanisms can be generated.

2. The FDA needs to be more creative in clinical trial paradigms: The FDA has been under siege recently because of its approval, and subsequent withdrawal, of diabetes drugs. As a result of the Avandia and Actos incidents, any new drug to treat diabetes is now required to complete an outcome study (in which patients are studied while on drug for three years and events such as heart

attacks, strokes, etc., are measured) before it is reviewed by the FDA. Studies like these cost hundreds of millions of dollars; and, although patients are on the drug for three years, these studies take five years to complete, given their complexity. In order to discover a truly novel breakthrough drug for diabetes, the FDA needs to be flexible in its approval requirements. Given the epidemic nature of diabetes, the FDA needs to address the disease with the same urgency with which it attacks AIDS and cancer. It should use its approach to AIDS and cancer trials as a blueprint for its treatment of diabetes, perhaps even reducing the initial outcome study to a year instead of three-years. If this study is successful, the drug can be approved by the FDA with the proviso that a three-year study will begin immediately on approval.

3. Pharmaceutical companies need to rededicate themselves to this area of research—there is no doubt that R&D in this area is risky. Novel mechanisms are speculative to work on, and the clinical trials are difficult, particularly if you are trying to measure the impact of an experimental drug on diabetic complications like kidney disease or retina degradation. Furthermore, there is risk involved when it is unclear how difficult the regulatory pathway is going to be. But Big Pharma needs to rise to the challenge here because they are the ones whose experimental drug will prove or disprove much of the early hypothetical work coming from academia and the NIH.

The cost to society from the diabetes epidemic is going to be huge. Changes in how we approach the discovery of new antidiabetics have to occur now in order to have an impact in the next decade.

BACTERIAL INFECTIONS

On the face of things, working in antibacterial R&D should be a "no-brainer." After all, you can never fully eradicate any pathogen. The reason for this is bacterial resistance. If you get an infection and are treated with an effective antibiotic, it is likely that the drug you take will wipe out as much as 99% of your infection. This impact is sufficient for you to feel fine and to return to your daily routine. But the small percentage of bacteria that survive do so because it is resistant to this medication. A drug designed to kill them can be extremely effective, but a small subset will be unscathed because they are genetically resistant to the way the drug exerts its lethal effects. As a result, some bacteria survive, multiply over a finite period of time, and emerge as drug-resistant pathogens.

This field of research, therefore, should be eternal. Infectious diseases can't be "cured." Furthermore, doing clinical trials is relatively easy. Unlike R&D for conditions like heart disease or Alzheimer's disease where thousands of patients need to be treated for years to measure the true effectiveness of a new medicine, a new antibiotic is used acutely. Generally, patients are dosed for two to four weeks and a physician can measure via blood samples whether the bacteria have been eradicated. Thus, the need for long and expensive clinical trials doesn't exist in this therapeutic indication.

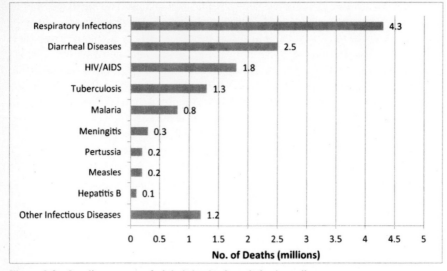

Figure 3.2   Leading causes of global deaths from infectious diseases.

The medical need in this area is undeniable. If one takes into account all infectious diseases, not just those caused by bacteria but also those viral in origin, it is estimated that they are responsible for 25% of global deaths (Figure 3.2).[22] The concerns around the growth of antibiotic-resistant organisms abound. Professor Peter Hawkey, chair of the United Kingdom's antibiotic-resistance working group, worries that resistant organisms are threatening to turn common infections into untreatable diseases.[23]

Why, then, are fewer and fewer large pharmaceutical companies investing in antibacterial R&D? Why are companies like Johnson & Johnson and Pfizer either scaling back their efforts or getting out of this field entirely? According to the Infectious Diseases Society of America (IDSA), of the 20 major companies doing R&D in this area 20 years ago, only two remain. How can this be? After all, according to the IDSA, antibiotic-resistant pathogens cost the US health-care system more than $20 billion per year. Yet, there are only a handful of new antibiotics currently in clinical trials, as opposed to the bounty of new drugs for cancer.

While there are a lot of attractions to doing R&D in this area, there are significant downsides. Although clinical trials for a new antibiotic are not daunting, the discovery of a new drug is very difficult. The resistant bacteria have evolved to such an extent that finding a new way to penetrate their defenses is much harder now than it was decades ago. Making modifications to existing classes of antibiotics is no longer fruitful; totally new molecular approaches are needed. Thus, just getting a new compound to test in the clinic may be more difficult in this therapeutic area than any other.

Just as important, however, is the fact that the commercial return on a new antibiotic would pale in comparison to a new treatment in areas like cancer or Alzheimer's disease. This is attributed to the fact that an antibiotic is generally used acutely, not chronically, and so profits are diminished by short-term use. However, a truly effective antibiotic against a life-threatening infection such as methicillin-resistant *Staphylococcus aureus* (MRSA) would likely receive very favorable pricing. The commercial issue really centers on the highly limited use of such a lifesaving antibiotic. In order to prevent overuse of such an important new medicine, federal agencies would limit its use so that it was the last line of defense. This would slow down the potential for drug resistance to the new antibiotic and prolong its effectiveness. This narrow use of the drug would limit its commercial viability, making it unattractive for a big company.

In 2011, the GAIN (Generating Antibiotic Incentives Now) Act was introduced by Senators Blumenthal (D, Connecticut) and Corker (R, Tennessee). A similar bill had already been introduced in the House earlier in the year. Its purpose is to incentivize companies to carry out antibacterial R&D. Among other things, the GAIN Act, when passed, will extend data exclusivity (the period of time before generic manufacturers can make use of valuable clinical trial data) for a new antibiotic by 5 years and streamline the FDA approval pathway and allow for rapid NDA review—welcome changes. But this reform probably isn't sufficient for the major companies to reenter this area. The GAIN Act should, however, be a boon to small companies that are seeking novel antibiotics.

In fact, this problem seems ideally suited to the expertise of biotech companies. An example comes from Optimer Pharmaceuticals. In 2011, Optimer received FDA approval for Dificid (fidaxomicin).[24] Dificid was designed to kill *Clostridium difficile*, a bacterium that can cause diarrhea so severe that it can be fatal. There are 14,000 deaths attributed to *C. difficile* annually. Sales estimates for Dificid for 2015 are $150 million, not significant for a large pharmaceutical company but a good product for Optimer.

Trius Therapeutics has a promising antibiotic in late stage development called tedizolid. This agent is active against serious resistant strains such as MRSA. The importance of tedizolid is that it appears to be effective against MRSA strains that have become resistant to Zyvox (linezolid), the current main treatment for MRSA.

Antibacterial R&D may be an area where the NIH can make a major contribution. Dr. Francis Collins, the Director of the NIH, has been seeking areas where the NIH can be more involved in quickening the drug discovery/development process. Perhaps using NIH resources to seek new ways to understand the genomics of resistant pathogens can lead to better insights into how best to kill them. More innovation is needed and the NIH can help fill this gap.

The IDSA has offered an interesting plan to help antibiotic R&D. They have proposed that the FDA treat new antibiotics as they do orphan drugs for rare diseases.[25] This would result in such drugs getting priority reviews by the FDA and NDAs based on smaller, focused clinical trials. The companies doing work on these antibiotics for hard to treat infections would get tax credits and other incentives. This is an excellent proposal that should be considered.

This problem is only going to get worse. It's great that Congress is taking steps to help work on this. Back in the 1980s, the FDA, academia, and pharmaceutical companies all worked together to discover new drugs that converted AIDS from a death sentence to a manageable disease. One would hope that it will not take a similar major disease outbreak to galvanize all parties to recommit to this field of research in a big way.

CONCLUSION

With regard to the productivity of biopharmaceutical R&D, I often hear it said that "all of the low-hanging fruit has been picked." The implication is that there once was a time when drug discovery was relatively easy and that treatments for hypertension, ulcers, or pain simply emanated routinely from research labs. It is as if people pine for the "good old days" of R&D as opposed to the current difficult times we face. Actually, memories of the way things were in R&D suffer from nostalgia, similar to the way that humans tend to focus on happy memories rather than the bad ones.

The fact of the matter is that drug discovery has always been hard and that there have been spectacular failures over the years, failures that consumed huge amount of resources. Back in the 1970s, one of the hottest fields in drug research was in the area of prostaglandins. These are naturally occurring compounds that play a role in a variety of biological processes including pain, vasodilation, and bronchoconstriction. There was a time when people believed that this field of research was as important as that of steroids and antibiotics. It was hoped that by developing prostaglandins or analogs, you would have new treatment for many diseases. Upjohn was the clear leader in this field, and there are tales that at one point they had as much as half of their research efforts devoted to prostaglandin work. Unfortunately, these molecules were difficult to synthesize, were not very selective drugs, and were short-lived in the bloodstream. A few compounds became products, such as alprostadil, an injectable treatment for erectile dysfunction, and misoprostil, to treat NSAID induced ulcers (which was discovered by Searle and not Upjohn). But these were relatively minor products easily superseded by other superior treatments. The only major product that came of all of this work was latanoprost for glaucoma. Overall, this area proved to be a huge disappointment, given the initial hopes and excitement that the discovery of prostaglandins generated.

In the 1980s, any company with an interest in drugs to treat asthma and COPD had a research program aimed at discovering an antagonist of platelet activating factor (PAF), a potent mediator of bronchoconstriction. There was good evidence that an inhibitor of PAF would be a useful treatment for asthma. PAF was found to be present in the lung fluid of asthmatics. In addition, administration of PAF to animals resulted in severe bronchoconstriction. At least a dozen companies found their own proprietary PAF antagonists and took these into the clinic. Surprisingly, not one was effective in asthmatics. PAF may be present in the lungs of these patients, but trying to block its action has no beneficial effect in asthma patients.

The pharmaceutical rage in the 1990s was in trying to find inhibitors of substance P. This is a neurotransmitter that was first isolated in 1931. It was of intense academic interest because there were hypotheses that substance P played a role in a variety of diseases such as pain, inflammation, depression, and even respiratory diseases. This field of research, however, was stymied by the lack of good substance

P antagonists, the tools needed to test the validity of these hypotheses. This research exploded when Pfizer announced the discovery of a novel class of potent substance P antagonists. Suddenly, a number of companies began to develop their own compounds, and dozens of compounds entered clinical trials. Unfortunately, none worked in the major diseases described. The one medical use of substance P antagonists was for preventing emesis. This is the mechanism of action for Merck's Emend, which has annual sales of $400 million—not nearly the type of blockbuster originally envisioned considering the early hopes for a substance P antagonist.

These are just a few examples, but I am sure that if you had a bunch of drug discoverers in a room, you would get their own favorite examples, such as aldose reductase inhibitors for diabetic complications or 5-lipoxygenase inhibitors for asthma and inflammation. If anything, all of these examples show that there are a lot of biological hypotheses that lie dormant until a biopharmaceutical company actually finds a compound that can be tested in the clinic to prove or disprove a specific hypothesis. But these examples also show that for decades the fruit has always been hard to reach. It is just that these days, the trees are higher due to the expectations of patients, physicians, and payers, as well as the existence of medicines of value to treat ulcers, hypertension, and high cholesterol.

Yet, there are still major medical needs that exist for cancer, Alzheimer's disease, atherosclerosis, diabetes, and resistant infections. There are also a host of other needs in orphan diseases, diseases of the emerging world, and even for areas where drugs already exist but where improvements are needed, such as depression and pain. To solve these problems, we need a vibrant and highly productive biopharmaceutical R&D industry. There are many "experts" who have suggestions on how best to improve R&D productivity. Are these the answers? This will be tackled in Chapter 4.

REFERENCES

1. Schmidt, C. (2011) Debate re-ignites contribution of public research to drug development. *Nature Biotechnology*, **29**, 469–470.
2. Stevens, A. J., Jenson, J. J., Wyller, K., Kilgore, P. C., Chatterjee, S., Rohrbaugh, M. L. (2011) The role of public-sector research in the discovery of drugs and vaccines. *New England Journal of Medicine*, **364**, 535–541.
3. Davis, P. B. (2011) Therapy for cystic fibrosis—The end of the beginning. *New England Journal of Medicine*, **365**, 1734–1735.
4. Boseley, S. (2011) Cancer research in "golden era," says charity chief. Guardian.CO.UK, August 22.
5. Hallberg, B., Palmer, R. H. (2011) Crizotinib—Latest champion in the cancer wars. *New England Journal of Medicine*, **363**, 1760–1762.
6. Eisenberg, C. (2011) NIH budget cuts may slow personalized cancer drugs, group says. Bloomberg, September 20.
7. Nelson, R. (2010) "War on cancer" more successful than perceived. *Medscape Oncology*, March 11.
8. Wang., S. S. (2011) Psychiatric drug use spreads. *Wall Street Journal*, November 16.
9. Mann, J. J. (2005) The medical management of depression. *New England Journal of Medicine*, **353**, 1819–1834.
10. Kramer, P. D. (2011) In defense of antidepressants. *New York Times*, July 9.

11. On August 6, 2012, Pfizer and J&J announced discontinuation of the bapineuzumab (intravenous) clinical trials due to lack of efficacy. Berkrot, B. (2012) Pfizer, J&J scrap Alzheimer's studies as drug fails. Reuters, August 7.

12. Belluck, P. (2011) Grasping for any way to prevent Alzheimer's. New York Times, July 20.

13. Peterson, E. D., Gaziano, J. M. (2011) Cardiology in 2011—Amazing opportunities, huge challenges. Journal of the American Medical Association, 306, 2156–2157.

14. Steinberg, D., Witztum, J. (2009) Inhibition of PCSK9: A powerful weapon for achieving ideal LDL cholesterol levels. Proceedings of the National Academy of Sciences, 106, 9546–9547.

15. Langreth, R. J. (2011) Heart attack stopping gene spurs biotech race. Bloomberg, November 11.

16. Cannon, C. P. (2011) High-density lipoprotein cholesterol as the Holy Grail. Journal of the American Medical Association, 306, 2153–2155.

17. Inazu, A., Brown, M. L., Hesler, C. B., Agellon, L. B., Koizumi, J., Takata, K., Maruhama, Y., Mabuchi, H., Tall, A. R., (1990) Increased high-density lipoprotein levels caused by a common cholesteryl-ester transfer protein gene mutation. New England Journal of Medicine, 323, 1234–1238.

18. Barter, P. J., Caufield, M., Eriksson, M., Grundy, S. M., Kastelein, J. J. P., Komajda, M., Lopez-Sendon, J., Mosca L., Tarde, J. C., Waters, D. D., Shear, C. L., Revkin, J. H., Buhr, K. A., Fisher, M. R., Tall, A. R., Brewer, B. (2007) Effects of torcetrapib in patients at high risk for coronary events. New England Journal of Medicine, 357, 2109–2122.

19. Danaei, G., Finucane, M. M., Lu, Y., Singh, G. M., Conan, M. J., Paciorek, C. J., Lin, J. K., Farzadfar, F., Khang, Y.-M., Stevens, G. A., Rao, M., Ali, M. K., Riley, L. M., Robinson, C. A., Ezzati, M. (2011) National, regional, and global trends in fasting plasma glucose and diabetes prevalence since 1980: Systemic analysis of health examination survey and epidemiological studies with 370 country-years and 2.7 million participants. The Lancet, DOI: 10.1016/50140–6736 (11) 60679-X.

20. Naik, G. (2011) Diabetes cases double to 347 million. Wall Street Journal, June 27.

21. Pollack, A. (2011) Novel diabetes drug is rejected by an FDA panel. New York Times, July 20.

22. Fauci, A. S., Morens, D. M. (2012) The perpetual challenge of infectious diseases. New England Journal of Medicine, 366, 454–461.

23. Laurence, J. (2012) Experts fear diseases "impossible to treat." The Independent, February 20.

24. Timmerman, L. (2012) Optimer, following Pfizer's playbook, has big plans for antibiotic. www.xconomy.com, January 31.

25. Yukhananov, A. (2012) Groups asks FDA to treat superbugs like rare diseases. Reuters, March 7.

CHAPTER 4

IMPROVING R&D OUTPUT

> Maybe the answer is for Pfizer not to worry so much about its own pipeline. Spending billions of dollars in R&D doesn't change the fact that the odds of any single compound making it to market are long, indeed. Perhaps it makes more sense, over the long run, to save on R&D and wait to license or acquire drugs (or companies) once they have made it to market or shown strong enough signs of efficacy and safety to de-risk the proposition.
>
> —Stephen D. Simpson, *Investopedia*, November 2, 2011

The pharmaceutical industry prides itself on its ability to discover and develop new medicines. Yet, this capability is being challenged because the industry's productivity has not lived up to expectations in the past decade. To someone like me, who spent 30 years in pharmaceutical R&D, the thought of totally outsourcing R&D is preposterous. But I suspect that Simpson isn't alone in his views. His challenge deserves an answer. Should research-based pharmaceutical companies forgo R&D altogether?

Before getting to the specifics of Simpson's proposal, it is important to understand the nature of a company's R&D budget line. First of all, everything is included in that figure, not just internal R&D costs but also things like the expenses generated in running co-development programs for clinical candidates partnered with other companies and milestone payments made to small companies whose early stage candidate it has licensed. Second, the size of an R&D budget is largely driven by "D," not "R." Based on my personal experience, the research component of a pharmaceutical company's budget tends to be only about 15% of the total as a result of the tremendous costs involved in developing a clinical candidate from Phase 1 through NDA approval.

In addressing Simpson's proposition, it must be noted that companies are already outsourcing a significant part of their pipeline. Martin Mackay, President of AstraZeneca's R&D division, has recently said that their goal is to have 40% of its pipeline generated from in-licensed compounds. My sense is that this is a goal shared by a majority of pharmaceutical companies, and, as will be discussed later, I support this strategy. But should a company be totally dependent on outside sources for its future?

Devalued and Distrusted: Can the Pharmaceutical Industry Restore Its Broken Image?
First Edition. John L. LaMattina.
© 2013 John Wiley & Sons, Inc. Published 2013 by John Wiley & Sons, Inc.

Simpson's comment that companies would be better off acquiring drugs that have shown enough signs of efficacy and safety to de-risk the proposition is, frankly, naïve. A compound can show promising signs of efficacy in Phase 2 studies, and you can license it at that point. However, you will still need to invest heavily in the Phase 3 programs needed to get the drug approved. Given the needs these days to perform very costly outcomes and differentiation studies, the bulk of the development costs remain in this aspect of Simpson's strategy. Furthermore, a program is never totally de-risked. There are a number of examples of drugs that prove to be disappointing commercially because of adverse events found post-launch when the drug gets used by millions of people.

So perhaps then, Big Pharma should forgo the licensing of compounds and simply buy companies, as Simpson suggests. History teaches that this strategy has its issues. Beyond the internal disruption not just of your entire organization but also of the acquired company, these mergers have not been viewed as being financially attractive by Wall Street analysts due to the premium that the acquiring company must pay and the lack of sustainability of the merged pipeline to meet long-term growth targets. Besides, you cannot just assume that the smaller company will automatically roll over and let itself be purchased, nor can you assume that smaller companies will be available on demand.

While Simpson didn't directly address the complete shutdown of a company's early research activities, others have. I also find this problematic. As noted above, the "R" component makes up only a lesser percentage of the overall R&D budget. And, when internal efforts deliver new products, you own them completely—no milestone payments, no royalties, no co-marketing deals. Thus, these products are more profitable. Even Simpson acknowledged that "Pfizer has some encouraging drugs that should come out soon," and most of these are internally derived. But it is also important to have a strong internal cadre of scientists to help evaluate the strength of the supporting data of compounds being considered for in-licensing. A good research organization provides this. Many a deal has been squashed by internal scientists based on their rigorous reviews.

Any company forgoing internal R&D risks its future. It is not something I would recommend.

This is but one example of a number of "solutions" to fix pharmaceutical R&D. This is a topic of interest for dozens of people: industry consultants, former R&D executives, Wall Street analysts, journalists, and even former FDA commissioners. It seems that everyone has the answer. Some of these proposals merit interest, but many have holes. This chapter seeks to review these ideas and critically evaluate them.

THE VIEWS OF OTHERS

Pharma's Blockbuster Mentality Needs to Change

Over the past few years, a number of critics of biopharmaceutical companies have predicted the demise of the industry because of its dependence on blockbusters. A blockbuster is defined as a branded prescription drug that generates annual revenues of $1 billion or more. Discovering a blockbuster should be a good thing because it is a medicine that is prescribed to millions of people because of its beneficial effects on disease and suffering. However, many major blockbusters, like Zyprexa, Lipitor, and Plavix, have already lost or are about to lose their patent protection and it is thought that, in the drug makers' pipelines, there is a dearth of new compounds with blockbuster potential to take the place of older products.

A few years ago, no less an authority than the former head of the FDA, Dr. David Kessler, slammed the blockbuster mentality saying:

> The model that we've based pharmaceutical development on the past ten years is simply not sustainable. The notion that there are going to be drugs that millions of people can take safely, the whole notion of the blockbuster, is what has gotten us into trouble.

Melody Petersen was even more strident in an opinion piece entitled "A Bitter Pill for Big Pharma"[1]:

> For 25 years, the drug industry has imitated the basic business model of Hollywood. Pharmaceutical executives, like movie moguls, have focused on creating blockbusters. They introduce products that they hope will appeal to the masses, and then they promote them like mad.

It's hard for me to envision my old boss, former Pfizer CEO Hank McKinnell, "taking a lunch" to discuss strategy with the heads of Paramount and Twentieth Century Fox.

First, it must be pointed out that a company doesn't set its research priorities based on whether or not a program can eventually yield a blockbuster. Such predictions are difficult, if not impossible. For a new medicine to be successful, it must be safe and effective and must meet a major medical need. Assuming that 15 years after starting a new R&D program, the new compound finally gets approved, it then needs to get a favorable label from the FDA, reasonable pricing from those who reimburse drug costs, and acceptance by physicians and patients. A great example in the difficulty of predicting blockbusters, interestingly enough, is the biggest blockbuster in history—Lipitor. When Warner-Lambert was seeking a partner to help sell and market what proved to be the biggest selling drug of all time, the company approached Pfizer. The Pfizer marketing team's analysis said that the peak sales potential of this medicine could be substantial. However, the actual peak in worldwide sales for Lipitor was almost $13 billion—more than anyone imagined. What happened?

The answer lies in the studies that Pfizer carried out with Lipitor AFTER it had already been approved and on the market. Pfizer invested over $800 million to show the importance of driving LDL cholesterol as low as possible. One such study was "Treating to New Targets" (also known as TNT). Conceptually, it was a simple study. Ten thousand patients with stable coronary artery disease and a baseline level of 130 mg/dL of LDL (which was considered reasonable 15 years ago) were randomly assigned to get either 10 mg or 80 mg of Lipitor and followed for nearly 5 years. At the end of the study, those on 10 mg of Lipitor had a median LDL level of 101 mg/dL and those on 80 mg had a median LDL cholesterol of 77 mg/dL. More importantly, those on 80 mg of Lipitor had 22% fewer heart attacks and 25% fewer strokes.

This proved to be a landmark study. For the first time, it was shown that lower LDL is better and that for people with a high risk of cardiac events, driving LDL levels down can be lifesaving. Suddenly, Lipitor's potency advantage proved to have a major clinical benefit. Pfizer also performed another major study known as CARDS (Collaborative Atorvastatin Diabetes Study), which for the first time showed that diabetics can reduce their risk of heart attacks and strokes by lowering their LDL levels with Lipitor. Similarly, in the lipid-lowering arm of the Anglo-Scandinavian Cardiac Outcomes Trial (ASCOT), lowering LDL cholesterol in patients with high blood pressure was shown to lower the risk of adverse cardiac events over hypertension therapy alone.

These studies and others helped to change medical practice. The importance of lowering LDL cholesterol as much as possible in patients at risk of a heart attack or stroke was unquestionable. In addition, these studies greatly expanded the patient population for those who would benefit from Lipitor therapy. With these data in hand, the sales of Lipitor soared. There is no doubt that the results from these studies proved to be crucial in recognizing the full potential of this important medicine.

That Lipitor evolved into the biggest selling drug of all time was due to a "perfect storm" of great efficacy, excellent safety, and the growing realization of the need to lower bad cholesterol (LDL) more than had been previously recognized. To base a company's strategy on such a sequence of events is foolish.

Nevertheless, with the large number of blockbusters going off patent, stories abound on the topic of the biopharmaceutical industry's need to change its business model. One such article had the following line regarding Pfizer:

> The next step will be rebuilding the world's biggest drugmaker into a smaller, faster moving company that focuses on development of biological drugs and specialty medicines.[2]

Unfortunately, specialty medicine R&D programs do not necessarily move more quickly to NDA approval. As was described earlier, the outstanding work that led to the cystic fibrosis drug, ivacaftor, began in 1989, and this drug reached the market in 2012. Furthermore, given the time it takes to go from an idea to a marketed product, any company, regardless of size, has a pipeline rooted in experimental medicines discovered five to ten years ago. Suddenly shifting one's strategy to specialty medicines is not feasible.

In reality, biopharmaceutical companies have been exploring new medicines for diseases that, while not representing patient populations of hundreds of millions, nevertheless are major medical needs. Such drugs are now being called "niche blockbusters," a term coined by Jonathan Rockoff in the *Wall Street Journal* in a story about a newly approved lung cancer drug called crizotinib (sold as Xalkori).[3] The article details the history behind the discovery and development of this breakthrough medicine. As was described in Chapter 3, crizotinib is an ALK-inhibitor that targets the genetic abnormality that causes about 5% of new lung cancers that are diagnosed each year. The beauty of this drug lies in the fact that a patient newly diagnosed with lung cancer can undergo a genetic test to determine if his or her lung cancer is ALK-dependent. If it is, then the physician now has a drug that is almost guaranteed to work in this patient.

The crizotinib story is not unique. Roche also has recently launched vemurafenib, a targeted drug for melanoma. We are moving away from the days when the only drugs that an oncologist had to treat his or her patients with were general cytotoxic compounds that come with myriad toxicities. The rapid advances being made in understanding the genetic basis of disease have led to the discovery and development of new drugs like these. But pharmaceutical R&D has been moving into this direction for a decade. After all, drug discovery is not an overnight process, and the research programs that are yielding these breakthroughs were started in the 1990s.

Thus, I am surprised by some of the following statements in the media that appeared as a result of Rockoff's article: "Can targeted drugs save Big Pharma?" "Perhaps the pharmaceutical industry has come kicking and screaming (to this)." "New cancer pill gives hope, new strategy." "There has been a paradigm shift." There still remains the view that pharmaceutical companies are only interested in Lipitor-like blockbusters and that smaller commercial opportunities are disdained. Of course, every company would love to have a drug with enormous sales. But very significant commercial returns can be made with crizotinib that more than justify its clinical development.

A number of years ago, I was asked by an industry analyst if I felt that the inevitable fractionation of patient populations by genetic subtype would be a death knell for Big Pharma. His rationale was that diseases like lung cancer would be treated with agents specific to a particular mutation for a subtype of lung cancer. Designed for very few patients, this process would result in treatments that are far less commercially viable than a lipid-lowering agent designed to treat millions. I explained that most drugs that are broadly prescribed do not work in a significant percentage of patients. The current paradigm is for physicians to prescribe drugs and then monitor their patients to see if the drugs are working. Often, patients come back complaining that their drug hasn't helped, leading the physician to try something new. This overall process is costly, inefficient, and frustrating to all concerned. By having specific, targeted drugs, physicians will have the confidence that the new medicine will help their patients, the patients will have confidence that the pills they are taking will help them, and payers will have confidence that they are paying for a worthwhile treatment. In such a world, while the number of patients taking the drug is relatively smaller, better pricing for the drug should be obtainable due to improved drug effectiveness. Basically, you might not have a few $10 billion-selling drugs, but you're likely to have many billion dollar products.

This scenario is being borne out by crizotinib. In his article, Rockoff reported that market analysts are predicting peak sales of crizotinib to exceed $1 billion. This would put the drug in the top third in sales for all of Pfizer's portfolio. Some might argue that the only reason that crizotinib will be a commercial success is that, as a cancer drug, it can command premium pricing. Perhaps this is true. But hopefully, genetically targeted drugs will be developed for other polygenic diseases like depression, schizophrenia, migraine, and so on. These patient populations should be large enough to support niche-like products with a reasonable price.

Targeted drugs have been envisioned and sought for over a decade. The first wave of these is now hitting pharmacy shelves. This is great news for patients and physicians—and not too bad for the companies developing them either.

There are a few lessons in all of this. First of all, predicting commercial success for a new medicine is always difficult. The biopharmaceutical industry has always been surprised, both positively and negatively, by the performance of new drugs. That is not going to change. To say that you are going to focus only on specialty products can be a prescription for disaster. Such compounds can play a role in a company's portfolio of products, but this shouldn't be the main driver for companies with sales of $25–$60 billion.

Second, from a discovery research standpoint, the resources needed to come up with a new compound for clinical development differ little for a specialty product or a projected blockbuster. Admittedly, the development costs for a specialty product can be less, particularly for a so-called orphan disease for which there is no treatment and relatively few patients worldwide. But a big pharmaceutical company's portfolio should have a limited number of such approaches if it is to thrive.

Finally, there is a view that there are fewer and fewer opportunities for major blockbusters. I beg to differ. A truly effective and safe drug to cause weight loss would likely have sales in excess of Lipitor's. The challenge in this field is clearing the high regulatory hurdle that exists for such a compound. A new drug that can slow or reverse Alzheimer's disease would also have tremendous commercial potential because the incidence of this disease will surge over the coming years with the increasing life spans of people globally. Heart disease continues to be a problem, and the obesity/diabetes epidemic will likely reverse the progress made in this arena over the past decade. Will an agent that raises the good cholesterol, HDL, be a new breakthrough for treating cardiovascular disease? If yes, major blockbuster status will be achieved here as well. There are also other major medical needs awaiting new, effective treatments.

You don't build a business strategy on the hope of discovering $10 billion/year products. You *do* base it on having the most productive R&D organization possible. And this leads to my final point. Slimming down R&D isn't the answer. Rather focus, stability, and resources are required for an R&D organization to thrive.

Can "Predictive Innovation" Lead to Greater Success Rates?

Drug discovery and development is founded on the premise that one can predict whether a compound will have clinical efficacy in a disease based on in vitro experiments in cells and in vivo experiments in animals. Unfortunately, scientists have shown for years that, while these models have some value, they are not foolproof. It has been said that cancer has been cured many times in mice. The problem is that results in rodents don't often translate to humans. Human beings are more than a giant two-legged mouse. Human biology and disease pathology are quite complex and cannot be easily mimicked in artificial laboratory settings.

Thus, scientists continually strive to find ways to identify research programs with inherently increased chances of success. This is referred to as "predictive innovation." This was described by Anders Ekblom, head of Science and Technology Integration at AstraZeneca, in the following way[4]:

> How early can I know that the approach I'm taking will definitely turn into a drug that delivers exactly what I would like to see? A lot of the cost in today's drug development is the cost of failures. We are all trying to focus our energy on how we can get different technologies to better predict outcomes.

A valuable paper by scientists at Eli Lilly[5] goes into great detail on this topic with respect to target selection. They believe that "validated targets for drug discovery are now materializing rapidly," and the attractiveness of many new targets is enhanced by the availability of biomarkers and surrogate endpoints which enable researchers to prove very early in clinical trials whether the hypothesis behind the compound's activity actually is relevant in humans. This is an excellent way to proceed, and I believe that all biopharmaceutical companies are seeking this path.

However, this approach to drug discovery isn't a guarantee for success, as evidenced by the two examples given in the Lilly paper. The first target is PCSK-9, a new approach for lowering LDL that was discussed in Chapter 3. There is a great scientific rationale for why an entity that blocks PCSK-9 function should be very effective in lowering a patient's LDL cholesterol. I will be shocked if the current clinical candidates now in early development, such as the PCSK-9 antibody, didn't have the desired effects. But this approach will not be guaranteed of success based on early clinical data. Remember that the CETP inhibitor, torcetrapib, also had excellent supporting science to justify advancing this novel HDL elevator to phase 3 studies (Part 3). Torcetrapib's dramatic lipoprotein remodeling was known from the very first patients dosed. Only the phase 3 results showed that this activity was irrelevant when trying to reduce heart attacks and strokes. The PCSK-9 blocker approach is based on the well-known benefits of lowering LDL cholesterol, so in theory the clinical effects on patients should be beneficial.

The scientific precedence that justifies the study of PCSK-9 blockers is also its Achilles heel. First of all, long-term studies comparing a PCSK-9 blocker with a statin will be needed to convince physicians and patients that they should switch to the former compounds. But payers will be another story. Why will they want to reimburse payments for a new but expensive treatment to lower LDL cholesterol when generic statins are available? Because the early PCSK-9 inhibitors are biological, such as antibodies, they will not come cheaply. Thus, even if a PCSK-9 were to successfully navigate the entire gamut of phase 3 studies needed to justify FDA approval, it may be limited for use only in patients who have dangerously high LDL cholesterol levels not controlled by statins.

The second target advocated by the Lilly authors as having a higher likelihood of success is an exciting new approach to treating pain involving a voltage-gated sodium channel known as NaV1.7. There is a rare disorder found in people born with a genetic mutation that prevents them from making NaV1.7. As a consequence of this defect, these people are insensitive to pain from birth. They are otherwise normal. Based on this genetic information, in theory a chemical that blocked NaV1.7 could be a great agent to treat severe pain without other side effects. Furthermore, one could demonstrate this activity in early clinical trials in people, and so clinical proof-of-concept could be had without hundreds of millions of dollars of investment.

This sounds great, but we must remember that, clinical outcomes are difficult to predict. Take, for example, studies with another novel experimental pain treatment, tanezumab, an inhibitor of nerve growth factor (NGF). This is a protein that modulates pain through sensitization of neurons. Multiple studies in animal models of pain show that NGF can both cause and augment pain. Furthermore, blocking NGF alleviates pain. About a decade ago, scientists at Rinat, a biotech company since acquired by Pfizer, developed an antibody to NGF called tanezumab. Tanezumab worked extremely well in animal models of inflammation, and so an obvious path for clinical study was to treat painful osteoarthritis of the knee, a poorly served condition.

The initial results with tanezumab were extremely exciting. Patients, for whom pain medications no longer worked and others who had been recommended for a knee replacement, suddenly felt great. Given that the sole biological role for tanezumab was to bind to NGF and prevent its harmful effects, it was thought that this antibody would have a great safety profile because it was a targeted drug. However, patients exhibited worsening of their arthritis, and some scientists have theorized that the patients were feeling so good that their rejuvenated active lifestyle resulted in worsening of their arthritis. It turns out that complete elimination of pain in these patients may not be such a good thing because pain serves as a warning sign that damage is occurring. Whether or not this theory is true, the clinical trials with tanezumab and the entire class of NGF inhibitors in knee pain were halted until regulators could better understand this effect and its overall impact on the risk–benefit profile.

The story isn't over for this compound. This is an important enough area of research that the FDA convened a panel of outside experts to review the data. This Advisory Committee recommended that studies continue with this drug with the rationale that there are some pain conditions for which there are few, if any, options for the patient. For example, the risk–benefit profile for tanezumab may be highly favorable in treating cancer pain. Thus, the FDA has removed the hold and allowed clinical trials with tanezumab and other NGF inhibitors to resume.

Perhaps the NaV1.7 program will yield a tremendous pain reliever with minimal side effects. Then again, maybe it won't. Pharmaceutical R&D is a high-risk, high-reward enterprise. There are no easy pathways to getting a major new medicine approved. "Predictive innovation" is part of the evolution of the drug discovery–development process. It may even improve the odds of success. But it is not a panacea.

Would Royalties Make Scientists More Productive?

Dr. Peter Hirth, the CEO of the biotech company Plexxikon, has had a very productive career in the biotech industry. He's played a key role in the discovery of three successful drugs, including Recormin for anemia (sold by Roche), Sutent for kidney cancer (sold by Pfizer), and the recently approved Zelboraf for melanoma (to be sold by Roche). His stellar track record was the subject of an article in *Forbes*.[6]

His views on the problems with industry productivity can be summarized as follows:

> Hirth's serial success stands out vividly in a pharmaceutical industry that has for years suffered from a profound innovation drought. He says that large companies should learn from what Plexxikon has done with a staff of only 43, explaining that if he were running a big business like Pfizer, he would form small units of 40 or 50 researchers and fund them sparingly but promise them royalties on any drugs that succeeded. Productivity, he says, would go way up.

There are a few issues with this statement. First of all, in big companies like Pfizer, the drug discovery project teams do, in fact, contain 40–50 scientists. This was true in my day with the teams in disease areas such as osteoporosis, atherosclerosis, ophthalmology, and urology. Yes, there were larger teams in broad disease categories such as Neurosciences, but this area was made up of teams working on depression, schizophrenia, addiction, and so on. Breaking this zone down into its components shows that the 40–50 scientist concept still holds. During my Pfizer tenure, the one exception was in oncology, where we had a group in excess of 200. Contrary to the view that size is detrimental to productivity, this group was the most productive we had at the time as judged by the number of clinical candidates it produced relative to the group's size.

Joshua Boger, an excellent scientist in his own right and the former CEO of Vertex, recently said that, in his experience, the relative size of the respective discovery groups doesn't matter—it's having the right people. I absolutely agree with him.

But the bigger issue for me with Hirth's statement is the fact that the overall process of going from an idea to the discovery of a clinical candidate to the conversion of this candidate to an approved new medicine *does* take an army. It's amusing to read "Plexxikon created Zelboraf . . . (and) Roche helped the tiny biotech test it." This "help" undoubtedly involved literally hundreds of Roche scientists to develop a formulation that enabled Zelboraf to be tested in the clinic, synthesize Zelboraf in sufficiently large quantities for clinical testing, run the necessary toxicology studies in animals to show that the drug was safe, and carry out the full gamut of Phase 1, 2, and 3 trials to justify FDA approval. My guess is that if you created a list of all the people who were involved in the discovery and development of Zelboraf, it would number at least 500 people.

And that brings me to my final issue with Hirth's stance, that of royalties. There are often seminal discoveries made in the development of a new medicine that extend beyond the discovery laboratory. Many a drug program has been saved by a key observation in the clinic on a drug's activity, or a breakthrough new formulation that allows the drug to be suitable to be made into a pill or capsule, or a key toxicology study. For example, the first observation of the potential use of Viagra for erectile dysfunction was made by a nurse who was monitoring male volunteers in the first clinical test of this drug. Should she get a royalty? I would argue that such a royalty scheme is unworkable because, in my experience, you would have to grant royalties to at least a dozen people who have made a seminal contribution. Unless the royalty was minuscule, it would greatly cut into the revenues for the company.

It is my experience that scientists are an extremely highly motivated bunch. They are driven by making use of their scientific talents to discover and develop something that, if successful, could benefit millions of people around the globe. I am not sure that the potential for a royalty would cause them to work harder. They are already incredibly dedicated.

Will Drug Repositioning Help Fill the R&D Pipeline?

The term "drug repositioning" is used to describe efforts around looking for new therapeutic opportunities for existing drugs. This also goes by other names such as "drug reprofiling," "drug repurposing," or "reusable drugs." The concept behind this is based on the principle that it is possible to find new uses for drugs that have already been approved. A drug approved by the FDA for a specific condition has cleared all the major safety, efficacy, and pharmacokinetic issues. Any new use for the drug that may be uncovered can be developed relatively quickly and cheaply since, theoretically, the major hurdle for this repositioned drug would be a clinical trial for the new disease indication.

It is interesting that whenever researchers discuss the value in repositioning old drugs for new uses, they always use sildenafil (tradename Viagra) as their poster child. This drug was serendipitously discovered as an agent to treat erectile dysfunction (ED), even though that wasn't the specific use it was designed for. However, this discovery was not so accidental. Sildenafil was designed as a potent inhibitor of an enzyme known as PDE-5. The interest in PDE-5 inhibitors stemmed from the fact that inhibition of this enzyme should result in elevation of nitrous oxide (NO) in vascular tissue beds. NO is well known to be a vasodilator. Pfizer scientists hoped that by blocking PDE-5 in the heart vasculature, arteries would dilate and the result would be enhanced blood flow in patients with cardiac disease like congestive heart failure.

Sildenafil did, in fact, cause vasodilation. However, this vasodilation was first observed in the penis and not the heart. Instead of being a breakthough medication for heart disease, sildenafil became a major treatment for ED. So, yes, this was a biological consequence that was not initially envisioned. The key in all of this is that Pfizer scientists designed and synthesized a safe and effective PDE-5 inhibitor that could be tested in clinical trials to determine what the utility of such an agent would be. Sildenafil was, in fact, designed as a vasodilator. Its effects, however, were manifest in an organ other than the heart.

When a new mechanism is found to be effective in patients, a company's scientists often explore where else such an agent may be of use. Pfizer researchers were also interested in learning whether sildenafil would cause vasodilation in other parts of the body. One theory was that the small arteries in the lung might also be sensitive to sildenafil's effects. Patients with primary pulmonary hypertension (PHT) suffer from arterial constriction which is extremely debilitating, and people with this disorder have trouble breathing. Clinical trials showed that sildenafil was effective in treating PHT, and it is marketed for this condition as Revatio.

Another example is Pfizer's tofacitinib, an inhibitor of the enzyme JAK-3. This orally effective drug was initially designed to be used as an agent to prevent organ transplant rejection. However, when the impressive early clinical data first came in, researchers began to envision other uses for a drug that acted by this mechanism, including rheumatoid arthritis and psoriasis. Hopefully, tofacitinib will soon be available for rheumatoid arthritis patients.

These two examples illustrate a few important points. The first is that, while serendipity is always appreciated in any research program, for any pharmaceutical research program to be successful, you need to have a safe compound that targets a specific biological process. Once in the clinic, you may find that the mechanism for which the drug was originally designed does not prove to be the optimum use for the new drug (there is also a downside to this: sometimes the new mechanism may have a mechanistically related side effect that turns out to kill the drug). But you don't go blindly into clinical trials with the hope that a PDE-5 inhibitor might do something beneficial in people. Rather, you must connect the mechanism to a biological effect.

The second point is that when a mechanistically exciting drug shows beneficial effects in a disease, the news spreads rapidly throughout a research organization. Scientists will share these results and then hypothesize where else such a compound may be effective. This leads to many other experiments to explore the new exciting finding and, potentially, new uses for this drug in medicine.

Finally, as a part of any drug's preclinical testing paradigm, it is explored in perhaps 100 different assays, to confirm the compound's selectivity. Thus, if it were to possess other unexpected activities, this is likely to be found in advance of the first human studies.

Admittedly, I am skeptical of the potential for drugs that have been discovered and developed over the past decade or so to be repositioned. However, I think there might be potential for repositioning much older drugs. A great example of this is the thalidomide story.

In the late 1950s, thalidomide was marketed in Europe. This agent was prescribed to pregnant women to help overcome morning sickness and to help them sleep. Unfortunately, insufficient toxicology testing had been done on this drug before it was approved. Thalidomide is a teratogen, a compound that interferes with the growth of the fetus and causes severe birth defects. Thousands of babies were born with severe deformities because their mothers took this drug. Fortunately, thalidomide was never approved by the FDA.

However, thalidomide was first repositioned 50 years ago.[7]

In the 1960s Jacob Sheskin, while working at Hadassah Hospital at Hebrew University in Jerusalem, was trying to help one of his leprosy patents sleep. He found some thalidomide and, remembering that it had helped patients with psychological problems sleep, he tried it on one of his patients. Surprisingly, the response to thalidomide in this patient and others was dramatic. Within days, most of the symptoms of leprosy disappeared. In what is a great example of the appropriate use of a drug once its risk–benefit is understood, thalidomide has become the drug of choice to treat leprosy, albeit under conditions carefully controlled by physicians.

Thalidomide was repositioned yet again a few years ago. In investigating thalidomide's beneficial effects in treating leprosy, it was found that this drug helps to stimulate the immune system. In addition, it appears to have antiangiogenic effects on tumor cells. This led scientists to explore the use of thalidomide in multiple myeloma, and it proved to be effective. It was approved by the FDA for this condition in 2006.

Efforts to reposition older drugs are flourishing in academic laboratories. Dr. Atul Butte and colleagues at Stanford have developed technology that allows them to screen genomic databases rapidly in such a way that they can identify examples where a drug creates a change in gene activity opposite to the gene activity caused by a disease.[8] Such an observation with an older drug could allow them to identify new uses for it.

The work at Stanford is not unique. Academic researchers are exploring a wide variety of drug classes such as phenothiazines, β-adrenergic receptor blockers and nonsteroidal anti-inflammatory drugs across a wide swath of diseases.[9] Even the NIH has gotten into the act with a deal involving the exploration for new uses of dozens of compounds from major pharmaceutical companies. I firmly support these efforts to find new uses for old or discarded experimental medicines. Hopefully, this work will lead to a major breakthrough. But I am not convinced that this will be a major source of new products. Casting a broad net with the hope of finding something new will be very challenging. The thalidomide example is more the exception than the rule.

Consultants Don't Always Have the Facts

It is no surprise that pharmaceutical companies make use of consultants. Often, they have skills lacking in a company. It is always a good idea to have a set of fresh eyes looking at the challenges and issues you face. Sometimes their advice is of value, sometimes not. But their views certainly can serve to broaden the internal discussion you are having and the decisions you will make.

It is surprising, however, to read the comments of people who purportedly are knowledgeable about the biopharmaceutical industry but yet who are totally off-base with their views. A *Wall Street Journal* op-ed piece[10] by Dr. Scott Gottlieb is particularly striking in this regard. Dr. Gottlieb was FDA deputy commissioner from 2005 to 2009. According to this op-ed, he now consults with drug companies. Here are some of his views and my thoughts on them.

1. "Drug companies learned that the big medical care needs—like lowering cholesterol levels—were no longer good clinical problems to go after." Actually, major medical need still exist in primary care areas like obesity and diabetes. Furthermore, the biggest news that emerged for the 2012 American College of Cardiology meetings was on the aforementioned new class of cholesterol lowerers, PCSK-9 blockers. Many believe that cholesterol lowering is still an important area of research.

2. "Companies are now aiming many of their new drugs on more serious maladies like cancer and Alzheimer's." In fact, Pfizer and other companies established major oncology research efforts in the late 1980s. The same can be said for Alzheimer's research. Any company just entering these fields is now too late to the game.

3. "The right size of a research team is now said to be 20–40 people." It is interesting that someone with no experience (to my knowledge) in running a discovery R&D lab in industry can offer such an opinion. While the pros and cons of size again will be discussed shortly, this comment may be accurate for running a single project. But, for running a portfolio of projects in an area as complicated as cancer or Alzheimer's disease, you need a much larger team of biologists, chemists, pharmacokineticists, clinicians, and so on.

4. "[Crizotinib] went from lab bench to patient bedside in about 6 years, rather than the 10–15 years it usually takes a new drug to reach the market." As was discussed earlier, crizotinib is a great example of the value of drugs that are considered "personalized" medicine—agents targeted for a specific condition. However, the discovery program from which crizotinib emerged in reality began about 10 years before it reached the market as Xalkori. There is no doubt that the crizotinib development program benefited from being a targeted medicine. But the discovery and development of any drug is difficult.

Dr. Gottlieb has other errors in how drugs were discovered in his piece as well. This discussion is not meant to single out Dr. Gottlieb. He is not unique in making recommendations on how to improve R&D productivity without necessarily having a complete understanding of the R&D process. Jack Scannell is an analyst at Sanford C. Bernstein. He attributes part of the decline in R&D productivity to the industry's use of high-speed technologies to aid in the drug discovery process.[11]

> The majority of first-in-class drugs (33/50) had their origins in phenotypic screening or were derived by modifying natural substances that were already known to have some kind of biological action. In other words, the processes that the industry was trying to abandon proved to be more successful at delivering major innovation than the processes that the industry was trying to industrialize and optimize.

First of all, the industry hasn't abandoned any method for discovering new biologically active molecules. Drug industry scientists are very pragmatic. They will use any method, tool, or paradigm to discover a new medicine. However, different projects call for different solutions. A few examples will help illustrate this.

In the 1990s, Pfizer scientists looked for compounds that could act as a nicotine partial agonist, a compound that they believed would be useful in treating smoking addiction. They approached this project as Scannell described. They started with cytosine, an alkaloid found in plants and known to have modest binding potential to nicotine receptors. After a couple of years of work spent in making and testing modified versions of cytosine, they found varenicline (tradename: Chantix), an important medicine to help smokers quit. Also in that decade, Pfizer scientists discovered the blocker of the enzyme, phosphodiesterase$_5$ (PDE-5), in a similar fashion.

Starting with known, nonselective PDE inhibitors, the team was successful in its search and discovered sildenafil.

In both of these cases, chemists were able to start with a known molecule with limited biological potential and, using their insight and experience, design a successful new medicine. However, oftentimes there is no compound available that can be used as a starting point. In other words, project teams often have an interesting biological target for which there is no known small molecule that interacts with it. The lack of a starting point (called a "lead") became particularly problematic as the function of more and more genes were discovered in the 1990s. Work from the Human Genome Project was providing exciting new theories on how to treat various diseases; but, without having a lead compound that modulated these new targets, drug discovery was impossible because there was no logical place for a chemist to begin synthetic efforts.

The solution to this problem was found in high-throughput screening (HTS), a technology discovery by John Williams and Dennis Pereira at Pfizer. HTS is the type of "industrialization" that Scannell rails against. It is an empirical approach based on testing millions of different compounds in microarray format to try to find a "lead," a compound with some small degree of activity that the team can use as a starting point. Much like cytosine was a starting point for the Chantix discovery team, any team running an HTS hopes to get a similar "lead" to enable the start of the discovery testing process.

So, how successful has the discovery "industrialization" been? Well, considering that it has only been available for 20 years and that the drug discovery–development process takes 10–15 years to complete, the results have been remarkable. A terrific perspective in *Nature Reviews Drug Discovery*[12] provides a history of the impact that HTS has had on the drug discovery process. Importantly, the article lists 11 recently approved drugs that had their origin in HTS, including Januvia, Iressa, and Tarceva. Others await approval, such as Pfizer's tofacitinib for rheumatoid arthritis. Without the invention and application of HTS, these drugs for cancer, AIDS, diabetes, and so on, would never have been discovered.

The pipelines of pharmaceutical companies are now replete with exploratory medicines that had their origins in "industrialized" methodology. Many of these will become important medicines to treat diseases in areas of major medical need. These technologies are invaluable in discovering new medicines, but scientists don't rely solely on "industrialized" solutions. It is a pathway used when appropriate. But to refer to this type of research as a blind alley is wrong. And to blame this methodology as a key reason for a decrease in the industry's R&D productivity is absurd.

PERSONAL VIEWS

In 1998 I was named the global head of Discovery Research for Pfizer. It was a wonderful opportunity because I was able to interact with outstanding scientists at our laboratories in Groton, Connecticut; Sandwich, England; and Nagoya, Japan. Working with these people was very rewarding because I was able to share in their efforts in diverse research areas like oncology, arthritis, pain, AIDS, psychiatric diseases, asthma, diabetes, and atherosclerosis. But my goals included getting to know the colleagues in the commercial division. Thus, periodically, I would go to the Pfizer headquarters in New York City and meet with different leaders in that organization. I got to know people who taught me a lot about the challenges they faced in commercializing a new medicine and what it took for a compound to be a success. I learned from these interactions and developed friendships over the years that have lasted to this day. But the person whose comments had the biggest impact on me was the person who, at the time, was the head of Pfizer's business in Latin America and South America–Ian Read, now Pfizer's CEO.

Ian has always been noted for his candor. From my first conversations with him, he expressed strong views about early research. One of his biggest concerns was in trying to evaluate the output of the discovery organization. He correctly opined that you can't really appreciate a given year's achievements in discovery until 10–15 years later. While a new drug candidate that hits a particularly interesting experimental target to treat cancer may be found, it may be shown to be inactive in a clinical proof-of-concept study conducted 5 years in the future. Thus, the oncology team could be rewarded for discovering a first-in-class experimental drug that eventually proved to be a bust. This experience was completely foreign for people like Ian and those in his commercial organization. They had hard annual targets to achieve that were clearly measurable every December. If they achieved these concrete goals, they were appropriately rewarded. If not, they paid the consequences.

This point of view had never previously been put forward so forcefully to me. I always knew that the organization depended on discovery research: If we didn't produce, everyone suffered. The discovery output was taken on faith by Pfizer. They put their trust in us to produce a portfolio of experimental medicines that would serve the corporation's growth objectives for the foreseeable future. Pfizer's R&D head at the time, John Niblack, often emphasized to me the importance of "getting discovery right." The views of Read and Niblack made a big impression on me and provided the foundation for many of my beliefs about R&D, beliefs that are outlined below.

Discovery Must Focus on Productivity

It is not unusual to hear a statement like: "If you do great research, you'll get new drugs." On the surface, this seems like a pretty reasonable comment. But, unfortunately, it is not true. A scientist can spend years doing excellent laboratory experiments, without having any of them lead to a potential medicine. The work done in pharmaceutical labs is applied research designed to take observations made in research institutes, government labs, or academia and to convert these into exciting compounds for clinical testing. Focus on getting a drug candidate is essential. A focus only on interesting science may lead to interesting publications, but not necessarily a drug.

Also problematic is that there are no sure things in discovery research. There are a variety of examples in this book where new exciting approaches for breakthroughs in treating a host of diseases, despite great scientific rationale and genetic information, turn out not to work when tested in clinical trials. Thus, we had at Pfizer what we called a "shots on goal" philosophy.

"Shots on goal" is a term obviously derived from sports. In theory, you have a better chance to score goals in hockey by taking 20 shots as opposed to taking only 10. Of course, they need to be *good* shots and not just wild attempts taken blindly. Each shot is taken not just to get it on goal, but rather to get it in goal. Thus, it would be a bit strange to hear a hockey coach differentiate between "shots in goal" and "shots on goal." Every hockey player takes each shot fully intending to score.

The "shots on goal" philosophy in drug discovery emerged from the realization that, no matter how good your research tools are, animal models such as genetically modified mice are very limited and are not necessarily good predictors of beneficial activity in humans. It is virtually impossible to predict whether a new discovery drug candidate will be a success or failure. There are a number of reasons for this. Does the new compound have an inherent unforeseen toxicological effect? Does the new mechanism being studied have fewer beneficial effects in patients than expected? Oftentimes, the true effects of drugs are only learned when they are tested in people. The unraveling of the human genome was a great boon to discovery scientists; it yielded a wealth of hypotheses as to how to go about treating, or even curing, a variety of diseases. The problem, however, was that while many of these new targets looked very promising in the lab, it was unclear as to which would translate into effective therapeutic treatments for patients. In vitro tests and animal testing often provided tantalizing results, but no guarantees.

The "shots on goal" philosophy was applied in the Pfizer oncology labs in the late 1990s and early 2000s as a result of the explosion of new targets that were emerging. At that time, there was a plethora of new ideas to explore in pursuit of discovering new treatments for cancer; some ideas were based on how to prevent tumor cells from metastasizing, others were focused on preventing tumors from growing blood vessels so they would starve themselves and die, some were specific to blocking the processes that caused tumors to grow wildly, and still other approaches were even designed to help the immune system fight off this awful disease. All of these were exciting, viable ideas. However, it was impossible to believe that any one of these would be a "magic bullet" to cure cancer. Furthermore, it was impossible to determine which approaches would be superior to others. The decision was made to find good compounds based on each idea and then take these compounds into clinical trials to see which, if any, successfully treated cancer. To be successful in the fight against cancer, a number of strategies—or "shots on goal"—were needed.

The ones taking these shots were in the Pfizer cancer discovery group, which grew to over 200 people, making it one of the largest divisions in the company. Over a 10-year period, it produced a number of clinical candidates that explored over 20 of these novel ideas to treat cancer. While many of these compounds failed to improve the survival of cancer patients, a number of them proved to be very effective. Some of these compounds have already reached the market such as Tarceva, Xalkori, and Inlyta. In fact, it can be argued that Pfizer has one of the strongest oncology pipelines in the industry. Yet, at the start of every one of these programs, one could have rationalized why they might not be successful—or why that "shot" shouldn't have been taken.

One might think that the "shots on goal" method is a scattershot approach, which, at best, yields a lucky result and, at worse, bloats the industry and inflates drug prices. Actually, it is a valuable R&D philosophy that has its foundation in the belief that you can't pick winners without clinical data and that a product doesn't become "unstoppable" until its full efficacy and safety profile are understood. However, for this philosophy to be successful, you must be certain that the "shots" are compounds that have cleared stringent hurdles designed to rule out any with potential flaws that would later reveal themselves in the key clinical trials. Just as a hockey coach puts his team in a position to win by encouraging his players to bombard the goal with good shots, a research manager increases his team's chances for success by exploring multiple promising drug candidates in the clinic.

The "shots on goal" philosophy is one designed for success, meant to maximize the overall productivity of an R&D organization. Every "shot *on* goal" is made with the full intent of it being a "shot *in* goal." The business of successfully discovering and developing new medicines is an incredibly risky endeavor, one that requires a lot of attempts before a winner is found. Limiting your shots by assuming that you can predict winners may ultimately prove to be a flawed strategy.

Does Size Help or Hinder R&D Productivity?

Most people know Pfizer as the behemoth it has become. But for the first half of my 30-year tenure, it was a relatively small company in a fragmented pharmaceutical industry. People are impressed with Pfizer's 2011 sales of $67.4 billion and their R&D spend of $9.1 billion. Few realize that these same numbers in 1992 were $6.8 billion in sales with an R&D investment of $851 million (data from Pfizer Annual Reports). It is interesting to note that Lipitor, at its peak in 2006, had sales of $12.9 billion—almost double the sales of the entire company 14 years earlier.

The company of the mid-1990s was quite different from what has now emerged. As an R&D organization, we were severely constrained. We worked in a limited number of therapeutic areas, and even in those we were able to explore only a few different approaches at any given time. We always tried to work in areas of high likelihood of success; but, as was already discussed, these are not easy to predict. The ability to identify "winners" at the early discovery phase is not easy. In those days, not only were we limited in the number of projects that we worked on, but also these projects were thinly staffed, particularly in comparison with our larger competitors. I would often hear our scientists lament that on projects where we had six chemists, Merck had 20. My response was always the same: "Our six are better than their 20!" However, it would be disingenuous for me to say that I didn't secretly envy the effort that Merck was able to put on their programs. To use a baseball analogy, the Tampa Bay Rays can compete with the New York Yankees and Boston Red Sox with half the budget of those richer organizations. But I would guess that Tampa Bay would like to compete with a similar budget. I felt the same way.

This all changed in the late 1990s. Thanks to the introduction of high-valued medicines such as Zithromax, Norvasc, Zoloft, Diflucan, Cardura, and Viagra, Pfizer sales grew dramatically and the R&D budget grew in parallel with this explosion in revenues. The $851 million R&D budget of 1992 was over $4 billion in 1999 (Pfizer 2000 Annual Report). Suddenly, we were able to run multiple programs in a given therapeutic area and with a healthy number of scientists. This was a terrific opportunity. Unfortunately, as we saw in Chapter 2, the rules of the R&D game changed dramatically at about the same time. R&D has become more difficult and expensive to carry out in the twenty-first century. This has contributed to the view that size is detrimental to productivity. But, I believe that size is not necessarily the culprit. I believe that success is related to how one utilizes the resources that you are allocated. As was already shown, the Pfizer discovery cancer team had a great deal of success and this success continues. This group was also the biggest we had at that time. Its success was not just due to size. We had a talented and competitive group that worked across three research sites: Ann Arbor, Groton, and LaJolla. But size wasn't a detriment, it was a great asset.

While this group was large in Pfizer's R&D setting, some would consider that it was not big enough. Colin Goddard, the former CEO of Oncogene Sciences Institute, a biotech company that was focused on cancer drugs, once told me that that to go after such an important disease as cancer, a team of at least 400 scientists was needed. Clearly, one's view of size is dependent on one's perspective.

There is a standard litany of the negative aspects of size: Decision making is slow; there is an aversion to risk; true innovation is snuffed out by layers of management. There is an assumption that small start-up companies are creative and big firms are not. Actually, my observations don't support this. A small start-up company is very tightly managed by investors, especially venture capital firms. In these situations, every experiment and every dollar spent by the start-up has to be justified. As a result, some "off-the-wall" experiments may not even be tried. In a big R&D organization, there are enough projects going on that you can take some shots that are high in risk. "Skunk works" projects can easily thrive.

An essay on the Schumpeter blog in *The Economist* entitled "Big and Clever: Why Large Firms Are Often More Inventive than Small Ones"[13] makes a very good case for size. Citing Michael Mandel of the Progressive Policy Institute, three arguments are given for the benefits of size in today's economy:

1. Growth is driven by big ecosystems that need to be managed by a core company that has the scale and skills to provide technical leadership; in a large R&D organization, there can be advantages in having immediate access to resources that a small company at best has to gain externally.

2. Globalization puts more of a premium on size than ever before; new ideas and breakthroughs are emerging all over, and tapping into them can be critical to success.

3. Many of the most important challenges for innovators involve vast systems, such as health care; to make a serious change to a complex system, you have to be big.

The comments of Dr. Joshua Boger, alluded to earlier, resonate strongly with me. Dr. Boger has lived on both sides of the fence. His early career was spent in R&D at Merck. He left Merck to head up Vertex, where he grew this start-up to become a much admired biotech company. Here is what he had to say in an interview with *Xconomy*.[14]

> I am a devout believer that size per se has nothing to do with innovation or its absence. Bell Labs was highly innovative for decades as part of the world's largest company. Merck was highly innovative from about 1950 to 1990. The key is culture, and you can throw away an innovative culture when you are small or you can drive it away when big. There are certain well-greased decline paths that larger companies often take that lead to stifling innovation, but these are not inevitable. . . . Once low emotional intelligence takes hold in the executive suite, value destruction follows. Too often shareholders and advisors ignore or deprioritize the kind of value-based culture in which innovation thrives. They over-control and over-measure and reward the wrong behaviors in favor of short-term objectives. This is the cause of innovation decline, not bigness.

My own beliefs are centered on building as strong a team of scientists as possible and set them off to build an optimal portfolio of projects. The size of the team is really dependent on the company's goals and objectives, along with the state of scientific knowledge in that field and whether it is ripe for investment. But whether you build a team of 200 in cancer or 40 in diabetes research, the principles are the same. Each team is empowered to create programs that will lead to exciting new candidates to meet a major medical need. Empowerment, however, doesn't mean abandonment. Periodically, it is important to review the progress that the teams are making and, if necessary, offer advice and adjustments. It is also important to be clear that the organization expects productivity. Projects that don't advance are closely scrutinized and possibly discontinued, in favor of more promising ones. But, if you have great scientists and you give them freedom and responsibility, they will be innovative regardless of how big the company is.

Properly used, size can be a great advantage. I would take that option any day.

To Outsource or Not to Outsource? That's the Pharma R&D Question

This question was posed by Sten Stovall of the *Wall Street Journal*.[15] Pharma companies are outsourcing more and more of the work it at one time did exclusively internally. At the start of Chapter 4, it was argued that relying totally on external sources for building one's pipeline is not a very sound business proposition. But does it make sense to outsource certain types of work? If yes, where can such a strategy be best leveraged?

Outsourcing can have three different purposes:

1. Budget sparing
2. Capacity generating
3. Pipeline building

The budget sparing situation is best exemplified by a conversation I had a number of years ago with the head of our R&D IT group at the time, who came in one day and said that he could outsource a large part of our IT work to a company in India at 10% of the cost of doing this work in the United States. He said that the quality of the work in India was excellent and that the work could be done in a timely fashion. How could any research head reject such an opportunity? Making such a move would enable us to reduce the IT budget and apply these funds to do added discovery or clinical studies.

The IT example is not unique. There are a number of processes in the drug discovery and development continuum that, while important, do not require in-house resources. Things as diverse as routine toxicology studies, preparation of synthetic intermediates, and the early-stage clinical studies can be done any place in the world. Such a strategy helps stretch the R&D budget as far as possible. One might ask if it makes sense to build your own capabilities to do this work in places like China and India. I would argue against this. It is likely that the low-cost locations of the future are not necessarily these but other countries. Maintaining flexibility in siting this type of work is key to maximizing your investment.

Historically, the pipeline of every Big Pharma company goes through times of plenty and times of relative drought. For this reason, you need to be able to outsource to generate enough capacity to move compounds rapidly through the different phases of development. It doesn't make sense to have the internal capacity that reflects your business needs for the plentiful times. Rather, you need to build an internal organization of modest size, and you must complement that with a network that you can tap into during times of peak productivity. This type of capacity generating outsourcing also helps to keep internal costs down.

But, perhaps the most important use of an outsourcing strategy is to build a company's pipeline. My view is that a pharmaceutical company should generate about 33% of its pipeline through outside sources (in-licensing). Actually, I didn't pull this number out of the blue. Rather, it is derived from some observations and beliefs that have been generated over the past 30 years.

Why in-license anything at all?

1. No matter how big your R&D organization is, no matter how capable it is, and no matter how smart your scientists are, it is impossible for one organization to corner the market on all the good ideas that are being worked on across the world. You will miss opportunities. To avoid missing out, it is important to be able to have the capacity to add such programs if they become available. Some years ago, when novel antiulcer drugs like Zantac were in late development, Merck realized that these compounds would be important new medicines. Because it didn't have its own internal program, it licensed two important antiulcer medications: Pepcid (famotidine) and Prilosec (omeprazole). In fact, in the case of the latter, Merck did state-of-the-art toxicology studies to help get this compound approved, thereby showing the added value that can be delivered by a partner with a strong R&D capability.

2. You may have recognized that a potential new breakthrough existed in a hot therapeutic area; but despite committing a good deal of time and effort, your own internal efforts failed. If having a new medicine in this area makes strategic sense for your company, it would behoove you to try to license a

promising agent from a company looking for a partner. This is what happened when Warner-Lambert Parke-Davis signed a co-marketing agreement with Pfizer for Lipitor. The result proved historic.

3. Perhaps your own internal program has had some success and your compound is looking good in mid-stage clinical trials. However, a competitor's compound is 1–2 years ahead of your own. Furthermore, your own compound looks pretty similar to the competitor's in terms of its clinical profile. It might make business sense for the two companies to link-up both programs and focus everyone's energy on the leading candidate. Why would the lead company do such a deal? For one thing, should a problem crop up with the lead compound, the second compound can serve as an alternate. In addition, the deal can be constructed in such a way that expensive clinical trial costs are shared. In addition, you would now have two shots in this arena instead of one. This was the situation when BMS sought a partner for its anticlotting agent, apixaban. While Pfizer had its own compound in development, apixaban was at least 12 months ahead and the Pfizer compound didn't have any apparent advantages. Joining forces maximized the use of resources for both companies.

This all sounds pretty good. Why not in-license 50–60% or more of your pipeline?

1. This is sort of like being dependent on foreign oil. Your pipeline is your life blood, your future. To depend on outside supply is a mistake. For one thing, you are never assured that what you need will be out there. Furthermore, you can't be guaranteed that you won't be outbid by someone else for what you want.

2. If you in-license a compound that makes it to market, in the best case you are paying a significant royalty. In the worst case, you are getting only 30–40% of sales. Clearly, you don't get as significant a return on investment for an in-licensed compound as one generated internally.

3. Most importantly, you need a significant in-house R&D group to be able to not just help evaluate potential in-licensed compounds, but also to be viewed by a prospective partner as an organization that will bring added value to its discovery.

One might argue with the 33% figure. Maybe it should be 30%. Maybe it should be 45%. The bottom line is that it should be a significant part of one's strategy. But it shouldn't be the majority. The pipeline is a company's lifeblood, and internal R&D must drive it.

Big Pharma is also making intense efforts to bolster its early discovery efforts by tapping into academic institutions to help convert early scientific findings into potential new medicines. This concept is not entirely new. In the past, companies often did collaborations with individual professors to gain access to a specific area of research. One of my former mentors, Professor Edward C. Taylor at Princeton, collaborated with Eli Lilly for many years, and the non-small cell lung cancer drug Alimta came from this collaboration. While I was at Pfizer, we set up a 5-year collaboration with the Scripps Research Institute in which Pfizer scientists openly collaborated with Scripps scientists in cutting-edge science programs.

However, companies are now building formal relationships with major institutions with the hope of broadening their research depth. This is yet another type of outsourcing, one intended to bring in new ideas and technologies at the earliest stage. Pfizer has set the pace in this area with their Centers for Therapeutic Innovation. The purpose of these units, according to Pfizer Senior Vice-President Jose Carlos Gutierrez-Ramos, is "to bridge the gap between scientific discovery and the delivery of promising compounds to the pipeline." Pfizer has established these centers with the University of California at San Francisco and the University of California at San Diego, as well as with major academic centers in Boston and New York. These are substantial relationships as shown with the UCSF interaction which, according to the Pfizer press release, could be worth potentially up to $85 million to the University.

Pfizer is not alone. Merck, Sanofi and other Big Pharma companies are investing in unique models designed to accelerate the translation of novel biomedical research into medicines for major medical needs. All of these initiatives are very exciting. The company–academic interactions between the scientists involved in these joint programs can be very stimulating and fulfilling. It is entirely likely that some great breakthroughs will result from these investments. There is only one caveat. The work being done is at the earliest stages of the R&D process. It will be a decade, at least, before any new product will emerge. However, this is a great step in improving R&D capabilities for both the companies and the institutions.

Big Pharma Early Research Collaborations

As was previously discussed, it is not unusual for Big Pharma companies to develop partnerships on late-stage development compounds where co-development makes great strategic sense for both companies and where such a collaboration can maximize the value of the emerging new drug. These types of deals have been done for decades. However, a new trend has emerged recently: companies pooling their resources to explore new research platforms or technologies in order to determine if there is value in the underlying science.

This type of interaction would have been shunned 5–10 years ago. The reason for this is that companies looked for any competitive edge in the discovery and development of new drug candidates. Any breakthrough was treated as a trade secret, and attempts were made to protect these new methods either by patents or "radio silence" in that the scientists didn't publish or talk about such research. An example of this was high-throughput screening (HTS). It was over a decade before Pfizer scientists published a seminal history of this work. By then, HTS was a common tool around the entire biopharmaceutical industry.

But, despite Pfizer's silence on the discovery of HTS, other companies realized that there was great value in being able to rapidly screen tens of thousands of compounds, and it wasn't long before this technology was broadly available. This shouldn't be surprising. The industry is full of bright, diligent, creative scientists who are always looking to make technology leaps to enhance the R&D process.

Thus, the thought that you have years of access to a new internally developed technique might be illusional.

In fact, new science has emerged over the past decade across the entire drug R&D continuum that offers the promise of improving this process, things like nanotechnology, biomarkers, imaging techniques, drug delivery technology, IT, and so on. The opportunities are such that it is impossible for one company to devote enough resources to explore any of these in real depth. As a result, companies are forming collaborations to explore such technologies and making any advances from this co-sponsored work available to all. This is a new attitude for Big Pharma companies, but it makes sense.

This is the rationale behind the creation of Enlight Biosciences, a company formed by PureTech Ventures where I am a Senior Partner. Enlight was created to meet "an unmet critical need in translating and funding the next generation of platform technology tools for drug discovery and development." The partners that are invested in Enlight are Merck, Lilly, Pfizer, AstraZeneca, Johnson & Johnson, Abbott, and Novo Nordisk. These companies select the areas of research, actively contribute to the technology development, and fund individual portfolio companies. A typical such company is Entrega, formed to address the issue of the delivery of potential drugs that are not absorbed in the intestine. The oral delivery of peptides, proteins, and even certain small molecules has been a major challenge for the biopharmaceutical industry for many years. Entrega has developed proprietary technology that may solve this problem. If successful, the pharma partners will have access to a revolutionary drug delivery methodology.

The work being done by Entrega would have been difficult to do in a single pharma company, because it would have diverted resources from ongoing internal programs. Furthermore, this is just one of multiple companies started by Enlight to explore novel science. No single company could pursue all of these in depth. However, through this consortium, the Enlight members can cast a wide net for the next technology breakthroughs. This is a terrific example of how precompetitive collaborations are helping to maximize the funds an R&D organization has to discover new drugs.

CONCLUSION

If you were reading this chapter with the hopes that you would find the answers to improving R&D productivity in the pharmaceutical industry, you've been sorely disappointed. There are no easy solutions. R&D is tougher now than ever before. New products need to be safer and more effective than existing medicines, and proving this takes time and money—a lot of both. Furthermore, the R&D process is still lengthy, even for so-called niche products that require small clinical trials.

Critics sometimes focus on the number of compounds approved as a way of measuring the industry's output. I am not sure that overall numbers are the issue. At the peak of industry productivity in the mid-1990s, the environment was very different. It was possible to have multiple viable products in a single disease category; statins are a great example. For reasons already described, that is no longer the case. In terms of overall benefits for patients, this is a good thing. We don't necessarily need six or seven entries in a single disease category; we would be better off with two or three drugs, each in different disease indications. Thus, we are all benefiting from this evolutionary thinking in R&D.

The bigger issue in terms of productivity is the cost now required to develop a new medicine. The hurdles set by payers and regulatory agencies, as well as the expectations of patients and physicians, require far more extensive data now than ever before in the history of R&D. This is not going to change. New biomarkers are not going to eliminate the need for outcome studies in drugs to be used chronically for the rest of a patient's life. Every drug, no matter how well researched, will have side effects. The question is what is the advantage that the new drug has in relation to its risk.

The biggest issue facing R&D organizations is trying to identify the fatal flaws of compounds before they enter the expensive late-stage clinical trial phase. One can be encouraged that new technologies already in hand or being developed will greatly help in this endeavor. However, measuring the impact on overall productivity takes time. Just as it took over a decade to see the benefits of high-speed technologies on discovery research, various new approaches such as the use of genetic information to improve clinical trial design or imaging technology to measure a drug's impact on disease progression are already being used as predictors for gauging the potential success of an experimental drug in late stage studies. It's hard to believe that these technologies won't have a beneficial impact on R&D output.

But there is another issue impacting R&D productivity that tends to be minimized. The cost-cutting culture of the past few years has had a tremendous unsettling effect on people in R&D organizations. It seems that everyone wants to shake things up. New R&D models are being imposed on organizations routinely. The threats of site closures and mergers have begun to have a numbing effect on researchers. In some cases, changes are being made before the last series of changes have been able to have a measurable impact.

The former head of Amgen's R&D organization, Dr. Roger Perlmutter, had some important insights on this. After his 11-year tenure as Amgen's leader, he was interviewed about his thoughts on the state of R&D in pharma.[16]

> My view of [R&D] is that yes, it's more expensive, and yes, it's harder than ever before, and the only way to succeed is to eschew distraction. It's so hard to do, to focus only on things that can make a real difference. . . . The challenge is to identify the critical important things that warrant your attention and throw yourself 100% into those things.

He is absolutely correct. To paraphrase a line made popular during the Clinton Administration about the economy: "It's the compounds, stupid!" This must be kept in mind when taking on any endeavor in R&D. Every investment, every organizational change, and every strategic shift should be made only after answering the following:

1. Will this improve the quantity and/or the quality of our candidate flow?
2. Is the disruption this will cause more than balanced out by the new value generated?

Researchers need not be coddled. In my experience, researchers will rise to the occasion to meet challenges the corporation faces. But CEOs need to realize that they can easily disrupt their company's engine. They need to help researchers focus on the task at hand—bringing new medicines to the world.

REFERENCES

1. Petersen, M. (2008) A bitter pill for Big Pharma. *Los Angeles Times*, January 27.
2. Armstrong, D. (2011) Pfizer after Lipitor slims down to push mini-blockbusters. Bloomberg, November 30.
3. Rockoff, J. (2011) Pfizer's future: A niche blockbuster. *Wall Street Journal*, August 30.
4. Ledford, H. (2012) Success through cooperation. *Nature News*, February 1.
5. Paul, S. M., Mytelka, D. S., Dunwiddie, C. T., Persinger, C. C., Munos, B. H., Lindborg, S. R., Schacht, A. L. (2010) How to improve R&D productivity: The pharmaceutical industry's grand challenge. *Nature Reviews Drug Discovery*, **9**, 203–214.
6. Herper, M. (2011) Serial lifesaver. *Forbes*, September 26.
7. Barber, S. (2007) Celgene: The pharmaceutical Phoenix. *Chemical Heritage Newsmagazine*, **24**, 1–2.
8. Marcus, A. D. (2011) Researchers show gains in finding reusable drugs. *Wall Street Journal*, August 18.
9. Oprea, T. I., Bauman, J. E., Bologa, C. G., Buranda, T., Chigaev, A., Edwards, B. S., Jarvik, J. W., Gresham, H. D., Haynes, M. K., Hjelle, B., Hromas, R., Hudson, L., Markenzie, D. A., Muller, C. Y., Reed, J. C., Simons, P. C., Smagley, Y., Strouse, J., Surviladze, Z, Thompson, T., Ursu, O., Waller, A., Wandinger-Ness, A., Winter, S. S., Wu, Y., Young, S. M., Larson, R. S., Wilman, C., Sklar, L. A. (2011) Drug repurposing from an academic perspective. *Drug Discovery Today: Therapeutic Strategies*, **8**, 61–69.
10. Gottlieb, S. (2011) Big Pharma's new business model. Drug makers aren't chasing blockbusters like Lipitor anymore, or uncovering compounds in the same way. *Wall Street Journal*, December 27.
11. Scannell, J. W., Blanckley, A., Boldon, H., Warrington, B. (2012) Diagnosing the decline in pharmaceutical R&D efficiency. *Nature Reviews Drug Discovery*, **11**, 191–200.
12. McCarron, R., Banks, M. N., Bojanic, D., Burns, D. J., Cirovic, D. A., Garyantes, T., Green, D. V. S., Hertzberg, R. P., Janzen, W. P., Paslay, J. W., Schopfer, U., Sittampalam, G. S. (2011) Impact of high-throughput screening in biomedical research. *Nature Reviews Drug Discovery*, **10**, 188–195.
13. Schumpeter (2011) Big and clever—Why large firms are more inventive than small ones. *The Economist*, December 17.
14. Timmerman, L. (2011) The fall of Pfizer: How big is too big for Pharma innovation. *Xconomy*, August 29.
15. Stovall, S. (2012) To outsource or not to outsource? That's the Pharma R&D question. *Wall Street Journal, The Source Blog*, February 7.
16. Timmerman, L. (2012) Xconomist of the week: Roger Perlmutter's parting thoughts on Amgen, *Xconomy*, March 1.

CHAPTER 5

RESTORING PHARMA'S IMAGE

> The problem is that drug companies are more focused on developing the drugs
> with the greatest market potential than they are on developing truly innovative
> treatments that address critical health needs.
>
> —Dr. Els Torreele[1]

Dr. Torreele's criticism is not unique. It is part of the harangue often used in discussing the failings of Big Pharma. But what is unusual about it is that this comment appears in a *Wall Street Journal* debate entitled "Should Patents on Pharmaceuticals Be Extended to Encourage Innovation?"[1] Dr. Torreele is one who believes that Big Pharma is not very innovative and that giving this industry the opportunity to extend the life of patents that were granted on "minor changes to an existing drug" doesn't foster innovation and isn't justified. There have been numerous examples given in this book about the value that the pharmaceutical industry brings to society. But the negative views of this industry are so pervasive that they are raised in widely different discussions, including this one on the relationship between innovation and extending patent rights. Dr. Torreele's comment derails the intent of this debate by basically implying that Big Pharma brings little to the table in terms of innovation; thus, extending the patent life on their drugs adds little or no value to the discovery of future medicines.

It wasn't always this way. There was a time when the pharmaceutical industry was the world's most admired. The leader was Merck, which was deemed by *Fortune* magazine to be the "World's Most Admired Company" for seven straight years starting in 1987. As recently as 1997, three pharma companies were in the top 10 of Fortune's list including Merck (#3), Johnson & Johnson (#4), and Pfizer (#8). What happened? Headlines like these tell the story:

> Drugmakers have paid $8 billion in fraud fines in the last decade.
> —*USA Today*, March 7, 2012

> Big Pharma's shame: emerging markets bribery.
> —MSNBC, MSN.com, February 28, 2012

Devalued and Distrusted: Can the Pharmaceutical Industry Restore Its Broken Image?
First Edition. John L. LaMattina.
© 2013 John Wiley & Sons, Inc. Published 2013 by John Wiley & Sons, Inc.

Side effects may include lawsuits.

—*New York Times*, October 2, 2010

Negative stories like these have become commonplace, so much so that the thought of pharmaceutical companies being "most admired" would seem ludicrous to many. This metamorphosis from being the purveyor of good to just another money grubbing machine damages more than reputation. Rather, we are beginning to see business problems emerge in this industry that go beyond the steep fines being paid. In an article entitled "Corporate Reputation Management in the U.S. Pharmaceutical Industry," Goldstein and Doorley outline the importance of corporate reputation in this field.[2] These authors make the point that corporate reputation can play a role in the pricing of drugs thereby impacting profitability. With people like Dr. Torreele casting doubt on the value that Big Pharma brings, patients and payers will balk at taking or reimbursing new medicines. When inevitable side effects are found for a new drug, rather than weigh the positives in the drug's risk–benefit profile, people will attribute such findings to Big Pharma cutting corners and rushing unsafe medicines to market. It is not unusual to hear a politician say during a campaign speech: "Elect me and I will protect you against Wall Street, oil companies, and Big Pharma!" What a long fall from "Most Admired"!

How serious could this become? Concerned that the fines are not proving enough of a detriment to prevent the illegal detailing of drugs, the Justice Department and the FDA are considering a variety of more drastic ways to penalize offending companies.[3] One idea being discussed is taking away a company's patent rights as a condition of any settlement. Another idea is to limit business with Medicare, thereby reducing the company's sales. Changes like these would have a bigger effect on a company's bottom line than even the billion dollar fines now being imposed.

There are a variety of things that Big Pharma can do to restore its image. These are discussed below.

ILLEGAL DETAILING OF DRUGS

When a new drug is approved by the FDA, it is approved for a specific disease or condition. This allows a sales representative (rep) to have a discussion with a doctor about the value that the new medicine brings to a patient. However, a sales rep can *only* talk about the use of the drug for treating the disease for which it is approved. For example, when talking about a new drug to treat pain due to osteoarthritis, a sales rep can extol the benefits of the new treatment for the pain of osteoarthritis. If the sales rep believes that the drug may relieve the pain of migraines, or even if the sales rep has knowledge that clinical trials are being run by his or her company that show that the new drug does, in fact, have efficacy against migraines, this cannot be discussed until the FDA has reviewed the supporting data for use in migraines and has formally approved it for this use as well.

Sales reps, however, can be overly aggressive in selling their drugs, so much so that they will push to have doctors prescribe medicines for yet unapproved ("off-label") indications. While doctors are free to prescribe drugs for any use, pharmaceutical manufacturers can only promote the drug for its labeled indications. Many of the fines that have been levied against pharmaceutical companies recently for such illegal marketing practices include GSK ($3.0 billion), Pfizer ($2.3 billion), Abbott ($1.6 billion), and Lilly ($1.4 billion). This creates a lot of cynicism about the motives of the pharmaceutical industry. Dr. Jerome Avorn, a Harvard Professor of Medicine and critic of the industry, explained it this way: "It's about the money. When you're selling $1 billion a year or more of a drug, it's very tempting for a company to just ignore the traffic ticket and keep speeding."[4] When people read stories like these, they are certainly not going to believe the industry's claims about the cost of developing new drugs or about the importance of pharmaceutical R&D to solving health issues.

All companies now claim that they have fixed these issues with internal training, compliance programs, and procedures. Lilly CEO John Lechleiter has stated:

> That was a blemish for us. We don't ever want that to happen again. We put measures in place to assure that not only do we have the right intentions, but we have systems in place to support that.

Only time will tell if these changes will be effective. One would hope that these behaviors will not be tolerated and that violators will face immediate loss of their job if they engage in this practice.

PHARMACEUTICAL COMPANIES SHOULD DROP TV ADS

Super Bowl commercials have become an integral part of pop culture, often generating as many stories as the game itself. Many ads have been proven to be extremely entertaining, so much so that they are viewed millions of times on YouTube both before and after the game. Drug company ads, however, have never shared such popularity.

In fact, drug ads seem to be more notable because of the controversy they cause. The most recent example surrounds celebrity chef Paula Deen. Ms. Deen is famous for her southern-style comfort food creations, dishes that, while tasty, rarely make the recommended menus of the American Heart Association. Unfortunately, Ms. Deen has been diagnosed with adult-onset type 2 diabetes, a disease associated with poor diet, lack of exercise, and obesity. Soon after she announced that she had this disease, Ms. Deen signed a lucrative deal to become a celebrity spokesperson for Novo Nordisk's diabetes drug, Victoza. This event led to renewed attacks against drug advertising. The following comments from Dr. Howard Brody, which appeared in *The Scientist*,[5] are typical:

> So, bottom line: is there something especially bad about any single celebrity deciding to shill for a particular drug or medical device, like Paula Deen telling us to eat cheeseburgers and also take good care of our diabetes? Maybe yes, maybe no. Is there a problem with how these products are marketed in the United States today? Absolutely.

In fairness, while I was a part of Big Pharma, I was sympathetic toward these ads. The justification for direct-to-consumer advertising has been focused on patient education. The stated goal has been to provide patients with information about new medicines and treatments for diseases that were previously untreatable. Furthermore, it is believed that advertising encourages patients to open a dialogue with their doctors about medical conditions and illnesses—communication that might not have previously existed. An example of how advertising can be beneficial is in the treatment of fibromyalgia. There was no medicine available to treat patients with this long-recognized painful condition. However, the FDA has recently approved drugs that work well. The companies that discovered these drugs have been able, via TV ads, to inform patients that there is now a treatment and encourage them to ask their doctors if they were a candidate for these breakthrough medicines. In such a situation, everyone wins: Doctors are able to alleviate their patients' pain, and the innovator company has a successful new medicine.

But the negatives that have evolved from TV ads are starting to outweigh the intended benefits. First of all, many consumers find the commercials offensive, pointing specifically to ads for erectile dysfunction. As a result, the FDA requires that, in terms of content and placement, television advertisements should be targeted to avoid audiences that are not age-appropriate for the messages involved. I am not sure that this is the type of aura that the industry wants around its image. Second, the intent and implication of these ads has come under particular scrutiny recently. A few years ago, Pfizer used Dr. Robert Jarvik, the inventor of the artificial heart, as a Lipitor spokesperson. Dr. Jarvik was, in fact, a Lipitor user. But critics attacked Pfizer for using Dr. Jarvik as an advocate for Lipitor, because it was felt that consumers would mistakenly believe that Dr. Jarvik is a cardiologist. In fact, although he has a medical degree, he has never practiced medicine. One could debate Dr. Jarvik's credentials to endorse Lipitor. On the one hand, he invented the artificial heart, so he clearly is knowledgeable about heart physiology even though he is not a practicing physician. But is it really worth the time and effort to do so? Did the negative publicity that arose from this incident justify having the TV ad? Probably not.

Another issue with these ads is the litany of side effects that a manufacturer must disclose in order to comply with FDA advertising guidelines. After listening to all of the potential toleration issues that one may get from the drug, it is a wonder that anyone would want to try it. Does hearing that a drug may cause "anal leakage" encourage you to take it? But there is another problem with the disclosure of side effects as was recently pointed out by Elisabeth Rosenthal in her *New York Times* article entitled "I Disclose . . . Nothing":[6]

When the Food and Drug Administration in the 1990s first mandated that drug makers list medicines' side effects in order to advertise prescription drugs, there was a firestorm of protest from the industry. Now the litany of side effects that follows every promotion is so mind-numbing—drowsiness, insomnia, loss of appetite, weight gain—as to make the message meaningless.

Finally, people believe that companies spend billions of dollars on the TV ads, money that could be better spent on R&D. In fact, some have the mistaken belief that more money is spent on direct-to-consumer advertising than on R&D. For the record, according to Nielsen, TV ad spending by the pharmaceutical industry was $2.4 billion in 2011. The amount spent by the industry on R&D was at least 30 times that amount. Nevertheless, this misconception only serves to continue the negative view that the public has regarding these ads.

If the pharmaceutical industry is really concerned about being better valued by the public, it might do well to drop TV ads completely. However well-intended they are, the negatives have always outweighed the benefits. If the members of the Pharmaceutical Research and Manufacturers Association agreed to halt TV ads, my guess is that the public's response would be overwhelmingly positive. My sense is that they wouldn't miss the commercials either.

THE NEED FOR GREATER TRANSPARENCY

In Chapter 1, the issue of transparency with regard to clinical trial outcomes was discussed. While the pharmaceutical industry is making progress in terms of registering clinical trials on www.ClinicalTrials.gov, it still has a way to go to meet expectations for prompt data entry. At a time when the industry is trying to rebuild trust with patients and physicians, the need to redouble efforts to get these studies available on line is obvious. Any perception that the industry is highlighting data that helps sell their drugs, while hiding less flattering data, encourages critics and counteracts any arguments in the industry's defense.

There are other themes that suggest the industry is reluctant to share and/or gather important information on new drugs. At times, a company is asked by the FDA to conduct studies after a drug is on the market to help better understand a potential emerging side effect. Such an example, on Merck's antidiabetes agents Januvia and Janumet, was recently reported by Ed Silverman on Forbes' Pharma & Healthcare blog.[7]

The main active ingredient for both of these compounds is sitagliptin, a mechanistically novel DPP-IV inhibitor that lowers blood sugars (Janumet is a combination of sitagliptin and metformin, a generic antidiabetic drug). The combined sales of both drugs was $4.7 billion in 2011. One concern that the FDA had with these compounds was the risk of acute pancreatitis, because they had received 88 reports of this condition attributed to sitagliptin between October 2006 and February 2009. To better understand this side effect, the FDA sent a letter to Merck requiring that Merck conduct a three-month pancreatic safety study in a diabetic rodent model. The letter was sent in October 2009. As of March 2012, Merck had not yet started the study. They are unlikely to start the study until late 2012.

Oftentimes, there are reasons for delays in starting studies like this. Sometimes, the drug sponsor needs to work with the FDA on the study design, because these usually aren't "cookbook" experiments. These studies need to be carried out in a way that both meet the scientific scrutiny of both the FDA and the Merck scientists so that both groups can feel that the results, no matter what the outcome, accurately reflect the safety profile of the drug. But to delay the start for more than three years after the FDA request is inexcusable. It is unlikely that the results of this study will cause the removal of these drugs from the market. They offer far too much benefit for patients. But the results of this study could cause the FDA to make a change to their recommendations on how the drugs should be used. Such a label change could impact sales. Critics of the industry point to a situation like this and cynically categorize this as intentional foot-dragging on the part of Merck to maximize profits for as long as possible. Their view is, once again, that the industry puts profits before patients.

The need for greater transparency also applies to sharing of data for the effectiveness of a drug. Tamiflu, an anti-influenza drug sold by Roche, was shown in clinical trials to reduce the risk of hospitalizations and complications from influenza. This was the basis of the New Drug Application (NDA) filed by Roche for the approval of Tamiflu. I have no doubt that the data showing the efficacy of Tamiflu is valid based on the FDA approval and the support of the use of Tamiflu by the Center for Disease Control (CDC). However, two researchers, Peter Doshi and Tom Jefferson, believe that Tamiflu is no more effective than aspirin in treating flu and they have made their views public in the *New York Times*.[8] In an op-ed piece entitled "Drug Data Shouldn't Be Kept Secret," they voiced their concern that, while the FDA may have reviewed the data, Roche has not published the data from eight of ten clinical trials involving more than 4000 patients, data that could support their views.

I must interject here that there often are claims that drugs are not as effective as manufacturers claim. The majority of such challenges tend to be based on incomplete understanding and/or faulty hypotheses from those making the challenges. But the lack of transparency in this case only fuels speculation and hurts Roche. The FDA, in my view, has pretty high hurdles when it comes to evaluation of safety and efficacy of new drugs. My guess is that the data that Roche provided for approval would stand up to the scrutiny of influenza experts around the world. Unfortunately, until Roche publishes this work, anyone could challenge that Tamiflu is no more effective than aspirin and, given the poor reputation of the industry, the public again will chalk it up to a pharmaceutical company foisting poorly effective drugs on patients and physicians.

Greater transparency is also needed in the industry–physician relationship. The pharmaceutical industry is extremely dependent on experts for advice on many R&D topics. These experts are in demand to lead or play a part in running clinical trials for new, important experimental drugs. These same experts are sought to comment on clinical design strategies or to provide advice on new breakthroughs in understanding diseases. These scientists/physicians are even sought to participate in mock advisory committee meetings that a company will conduct to prepare for the real FDA Advisory Committee hearing that guides approval.

When companies do this, they will always try to get the leaders in the field. These leaders can and do command high fees for their services—justifiably. This interaction is very important in discovering and developing new medicines.

But the transparency of these payments has not been rigorous. In fact, for years these payments had been largely unreported. This is another example of the industry getting a black eye for mishandling an important component of its work. When a leading physician supports the approval of a new drug, patients want to know how much money, if any, this consultant received from the drug's sponsor. As stated by Allan J. Coukell, a pharmacist and consumer advocate at the Pew Charitable Trust:

> Patients want to know they are getting treatment based on medical evidence, not a lunch or a financial relationship. They want to know if their doctor has a financial relationship with the pharmaceutical company, but they are often uncomfortable asking the doctor directly.[9]

As a result, the US government has had to take steps to mandate such transparency. As part of the new health-care bill, companies must report *all* payments to doctors which are made to help develop, assess, or promote new products. Hopefully, the industry will do a better job of adhering to these new rules than it's done with entering clinical trial data into public registries, which is now mandated under federal law.

The issue in all of these examples is that the industry could have avoided many of these problems. Getting data onto www.ClinicalTrials.gov is largely one of applying extra resources on the project and creating a sense of urgency. The same can be said for completing experiments or clinical trials that the FDA requests. And can the industry really be surprised by the outcry and the doubt created in its work when people hear in the newspapers that the support of key opinion leaders could be tainted by payments? These are the type of issues that the industry must address proactively if it is to regain public trust.

HOW COMMITTED IS BIG PHARMA TO RARE DISEASES?

Rare diseases are serious conditions that are at the very least disabling and at worst can be life-threatening. Generally defined by the National Institutes of Health (NIH) as conditions that afflict fewer than 200,000 individuals, there are close to 7000 such diseases. Some of the more common rare diseases include muscular dystrophy, cystic fibrosis, and multiple sclerosis, but there are others that impact less than a dozen people in the world.[10] The vast majority of rare diseases are genetic in origin and can be caused by a defect in a single gene or due to mutations in several genes.

Until recently, research into drugs to treat rare diseases had largely been the province of small biotech companies. Larger companies had avoided rare diseases to focus on medical needs that had bigger patient populations including cancer, Alzheimer's disease, and diabetes. These larger potential patient populations would theoretically translate to higher revenues for new, effective treatments. It was anticipated that, given the small patient populations for rare diseases, similar revenue streams were not possible. A drug that treated only 1000 patients couldn't possibly be as lucrative as one that treated 10 million.

However, biotech companies that discovered and developed drugs for rare diseases were able to charge very high prices for these lifesaving medicines, and these drugs became major revenue generators. A great example is Genzyme's Cerezyme for the treatment of Gaucher disease, a genetic disorder that impacts organ function. Cerezyme is extremely effective, but it is also extremely expensive with annual treatment costs as high as $300,000/patient. As a result of this high price, Cerezyme's 2010 sales were over $700 million, a respectable number for any drug. Genzyme's success in the rare disease area attracted the attention of Big Pharma. In fact, last year Genzyme itself was acquired by Sanofi. Companies like Pfizer and GlaxoSmithKline established their own research units exclusively devoted to seeking cures for rare diseases. Suddenly, drug research into rare diseases was not exclusive to biotechs.

This is a welcome development for patients and the families of patients with loved ones suffering from rare diseases. However, these people are also wary about the entry of Big Pharma into this arena. This concern was eloquently expressed recently by Melissa Hogan in her *Saving Case* blog. One of Ms. Hogan's sons, Case, suffers from mucopolysaccharidosis (MPS). MPS is caused by a genetic mutation that prevents a patient from producing iduronate sulfatase, an enzyme needed to cut up polysaccharides. Incomplete breakdown of polysaccharides causes these compounds to remain in cells, causing progressive damage both physical and mental in nature. Three of her four concerns, I think, are easily addressed:

1. *Transparency.* She asks that companies be open and honest with any success or lack of success in a given area; let the patients be among the first to know so that they don't find out information by stumbling onto it. I think this is easily done.

2. *Compassion.* "We want to know that you care about our children, that you care what happens to them as people without respect to whether you are able to sell them a product or not." Not only is this doable, but I think it would behoove these companies to have people like Ms. Hogan visit their research labs to tell their story. This sort of experience can be incredibly motivating for scientists. Having advocates see the commitment of scientists can be equally rewarding.

3. *Partnership.* Ms. Hogan's worry is that Big Pharma will barge in with the attitude that "we are here to help, but we know what is best." She is asking for Big Pharma to view the patients and their families as partners in the process. This request makes a lot of sense in that this group will be very helpful in being part of the clinical trials necessary to get a drug approved and ultimately accepted by physicians and payers.

However, it is Ms. Hogan's worry about the industry's commitment to rare diseases that I, too, believe is cause for concern.

We want to know, especially from Big Pharma, that some might see as interlopers seeking to bleed the last of high profits from a new area now that their more common drugs are going generic, that entry into this area is not a fleeting thought of profits. We want to know that if the research is more difficult, that if the science takes longer, that if the competition gets heavier, that you are committed. We want to see understanding that commitment is not just offering us a product, it is helping to make our children's lives better, a product being just part of that effort. Because it's not commitment to us, it is commitment to our children.

It is easy to understand Ms. Hogan's worry. Big Pharma was late to the table for rare diseases and, in reality, was led there by the financial success small companies are having with high priced products. I do not worry that Big Pharma will find that R&D on rare diseases will be more difficult or take longer than R&D for drugs to treat major diseases. In fact, one can make a case that R&D in rare diseases can be easier (fewer patients needed for clinical trials, trials of shorter duration). My worry stems from my belief that the current pricing for rare disease drugs is unsustainable. As more of these drugs become available, it will become difficult for governments and health-care providers to reimburse these costs. As a result, one can envision a scenario where the prices of these drugs are controlled (as already happens in areas outside the United States). Such an action will reduce the revenues that a company can expect with these drugs. If the profit margins shrink on these drugs, will Big Pharma continue to work on rare diseases? It must.

Big Pharma can gain a lot of good will, something it desperately needs, by fully committing to rare diseases in their internal and external behaviors. A company should tout its successes in this type of R&D, fully embrace the patient advocacy groups that work so hard on these diseases, and provide other types of patient support in areas where drug treatments are not available. In addition, it can make research centers that are working on rare diseases become a focal point for researchers around the world with seminars and lecture series. Such actions would be incredibly motivating not just for scientists but for all colleagues in the company. Not only would this be good for business, it would be good for a company's culture and morale.

Rare disease efforts should not be exploited—they should be embraced.

PHARMACEUTICAL COMPANIES AND PHILANTHROPY

Unfortunately, even when the pharmaceutical industry engages in selfless work, they don't seem to get a lot of credit for it. Early in 2012, at a major press conference, the Bill & Melinda Gates Foundation and 13 drugmakers, along with the World Bank, the United States, Great Britain, and the United Arab Emirates, announced a joint effort to attack 10 neglected tropical diseases. The drug companies will be contributing $785 million worth of drugs to this effort, and Gates is pledging $363 million to try to eliminate these diseases in the next decade. This is indeed a noble initiative, one that should be applauded for its ambition and scope. It will involve not just money and medicines, but also the talents of scientists at these companies who will help guide the critical research needed to make breakthroughs in eradicating these diseases. The fact that so many organizations are working together for the first time should help drive the success of this initiative.

But, for the pharmaceutical companies involved, this isn't something new.

For decades, pharmaceutical companies have been working on programs designed to help people in the developing world. In 1987, Merck began efforts to eradicate river blindness, a disease spread by black fly bites and characterized by disfiguring dermatitis and eye lesions leading to loss of sight. Merck formed a partnership with the World Bank, the WHO, UNICEF, and various ministries of health to provide free Mectizan (ivermectin), which treats river blindness with a single annual dose, to anyone who needs it. More than 69 million Mectizan treatments are given each year across 33 different African and South American countries. The WHO estimates that 40,000 cases are prevented annually as a result of this program.

The Merck example is not unique. Zithromax (azithromycin) is a great drug to treat *Chlamydia trachomatis*, the bacterium responsible for causing trachoma and the leading cause of preventable blindness in the world. In 1998, Pfizer co-founded the International Trachoma Initiative (ITI) and, through the ITI, has provided over 54 million treatments of Zithromax to trachoma patients in 15 countries. This program is part of the WHO's SAFE (Surgery, Antibiotics, Face-washing, and Environmental improvement) strategy, which is designed to eradicate trachoma by 2020.

Virtually, every major pharmaceutical company has been involved in these types of efforts. GSK, AstraZeneca, Lilly, Sanofi-Aventis, and Novartis are all working to combat tuberculosis. Similarly, Pfizer, GSK, Novartis, Eisai, and Sanofi-Aventis are all working on malaria. The same can be said for AIDS and tropical diseases. In fact, the International Federation of Pharmaceutical Manufacturers & Associations (IFPMA) lists on its website 213 different efforts aimed at improving the plight of those suffering in the developing world.[11]

Moreover, pharmaceutical companies historically have led all businesses in terms of generosity. In her *Forbes* article entitled "America's Most Generous Companies,"[12] Jacquelyn Smith reported that in 2009, according to the *Chronicle of Philanthropy*, six of the top 10 corporations were pharmaceutical companies. Pfizer led the list with $2.3 billion donated in total cash and product-giving. The top 10 also included Merck, Johnson & Johnson, Abbott, Lilly, and Bristol-Myers Squibb.

Cynics will say that pharmaceutical companies are only doing this to help their image. The fact is that the companies have been doing this for so long that it is part of their culture. The men and women in these companies who help in these types of projects take great pride in this work. This new public–private partnership to combat 10 neglected diseases is a terrific initiative and hopefully will be met with great success. But for pharmaceutical companies, it is a continuation of decades of work and billions of dollars of investments. Hopefully, the collaboration with the Gates Foundation and others will generate much needed visibility to Big Pharma's effort in this area.

PHARMA NEEDS TO HAVE ITS SCIENTISTS TELL THEIR STORIES

The general public is bombarded on a daily basis with negative stories about the pharmaceutical industry. Lawsuits on the side effects of new drugs, doctors publishing papers "ghost written" by pharmaceutical companies, and companies hyping new drugs that some claim are no better than placebo all contribute to the hole from which the industry must climb out. Pharmaceutical companies do tremendously valuable work and are a key component for solving the medical challenges facing patients around the world. But the cynicism that the industry faces runs deep. Ads touting the value that pharmaceutical companies bring are looked upon suspiciously and can generate a comment like "What unnecessary drug are they trying to foist on us now?" I once heard a critic say that the only reason that pharmaceutical companies are working on new treatments for cancer was so that drug companies could change their image and that, in reality, companies would rather work on lifestyle-enhancing drugs.

But companies have an untapped resource that can help turn their image around—their own scientists. I witnessed this myself in my days at Pfizer. Scientists are very passionate about what they do. One would have to be so dedicated, given that a scientist will likely have devoted years to a specific project seeking a new treatment for cancer, Alzheimer's disease, or a deadly infectious disease. When scientists describe their work, with the failed experiments, the new theories, and the first big breakthrough, followed by a new unanticipated hurdle and finally a successful new medicine, the audience gains a new appreciation for how hard an enterprise drug discovery and development truly is. When people hear these stories, they come away with an appreciation of the work that is being done. They also gain a great respect for the people in the lab who are dedicating themselves to finding cures.

At Pfizer we had created a "Science Ambassadors" program. Each ambassador, a veteran of R&D, was asked to give two to three talks per year to various organizations and stakeholders in order to teach about the work we were doing. They gave talks at places as diverse as university hospitals, state legislatures, health plan associations, patient advocacy groups, and local radio stations. The feedback from these sessions was always excellent, some of which is captured below.

> "This person is an outstanding spokesperson for your company and I believe that the future of anti-infective medicines is in good hands."

> "You can't fully discuss public policy issues around the pharmaceutical industry until you know what it is the industry does—innovative R&D."

> "Pfizer is doing amazing work . . . your scientist has demonstrated to one of the most respected academic medical centers that there is much more to the pharmaceutical industry in general and Pfizer in particular than had ever been considered."

Not only did audiences gain from these talks, the experience had a great effect on the scientists as well. They found the interactions energizing, and they loved being able to contribute to the challenge of enhancing the reputation of the industry in which they were dedicating their professional lives.

Unfortunately, you cannot attack this problem with just a handful of scientists from one company. This should be a major initiative led by the Pharmaceutical Research and Manufacturers Association (PhRMA). A coordinated effort with all companies involved could begin to help change how pharmaceutical companies are viewed. Wouldn't it be great to see an episode of *The Dr. Oz* show dedicated to the story of a new drug to treat breast cancer, as opposed to one focused on the side effect of a drug that occurs in 13 out of millions of treated patients?

CONCLUSION

A person who is not a fan of the pharmaceutical industry could read this chapter and not be moved. After all, the pharmaceutical industry got themselves into this mess.

They deserve what misfortune falls on them as a result of these behaviors. Well, perhaps that is true. Unfortunately, the lack of credibility that the industry is facing is negatively impacting medical practice. This can be seen in the story of Gardasil, expertly told by Matt Herper.[13]

In 1976, it was discovered by a German scientist, Harald zur Hausen, that cervical cancer tumors were caused by human papilloma virus (HPV), a discovery that won zur Hausen the 2008 Nobel Prize in Physiology & Medicine. Work at both the National Cancer Institute and later by Merck led to Gardasil, an excellent vaccine to treat HPV. This is such an important vaccine that the American College of Obstetricians & Gynecologists strongly recommends its use for 11- and 12-year-old girls. There was a time when this vaccine would simply have gone on to become a successful product and Merck would have been lauded for its scientific prowess and contribution to medicine. Unfortunately, the story doesn't end here. Merck began a major advertising campaign on the risks of HPV and lobbied state governments to make vaccination with Gardasil mandatory, a major error according to the former FDA head, Dr. David Kessler: "It would not have been the way the old Merck would have done business. It was a science based company, not a marketing company."

The outcry about mandatory vaccinations was huge and even became a topic in the Republican Presidential debates. Governor Rick Perry of Texas had endorsed mandatory vaccinations in his state and was attacked for this decision, somewhat irrationally, by another candidate, Representative Michele Bachmann. But there was no getting around the fact that Gardasil was a Merck product and that Merck was facing lawsuits on its arthritis drug, Vioxx. As a result, Merck's motives and credibility were already under siege, and the ultimate losers in this whole situation are patients who will get cervical cancer as a result of HPV. Thanks to the availability of Gardasil, this is a preventable disease. Yet, because of all the controversy surrounding this vaccine, it is estimated that only 25–30% of the targeted population is being immunized. This is terrible. Studies in thousands of patients have shown that it is a safe and effective vaccine. Its broad use will save lives. Unfortunately, the lack of confidence in the pharmaceutical industry is limiting its use.

Of course it is important for the pharmaceutical industry to regain its lost luster. Any business is helped by being viewed positively. But I would argue that a return to past respect for this industry would benefit patients as well. It is well known that drugs, when used properly, can provide great savings to the overall costs of health care. Yet, when patients have doubts about their medicines and the value that they can provide, they are apt not to take them. Restoring pharma's image can benefit us all.

REFERENCES

1. Bloom, J., Torreele, E. (2012) Should patents on pharmaceuticals be extended to encourage innovation? *Wall Street Journal*, January 23.
2. Goldstein, K., Doorley, J. (2011) Corporate reputation management in the U.S. pharmaceutical industry. *Institute for Public Relations*. www.instituteforpr.org.
3. Kennedy, K. (2012) Drugmakers have paid $8 billion in fraud fines. *USA Today*, March 5.
4. Wilson, D. (2010) When the side effects may include lawsuits. *New York Times*, October 3.

5. Brody, H. (2012) Opinion: Celebrities pushing drugs? Celebrity spokespeople for pharma companies can manipulate the public's understanding of disease. *The Scientist*, January 30 http://thescientist.com/daily/2012/01/31a.htm.

6. Rosenthal, E. (2012) I Disclose . . . Nothing. *New York Times*, January 22.

7. Silverman, E. (2012) Waiting For Godot: Merck The FDA And Post-Marketing Studies That Never Arrive http://www.forbes.com/edsilverman/2012/02/29/

8. Doshi, P., Jefferson, T. (2012) Drug data shouldn't be secret. *New York Times*, April 10.

9. Pear, R. (2012) U.S. to force drug firms to report money paid to doctors. *New York Times*, January 12.

10. Melnikova, I., (2012) Rare Diseases and Orphan Drugs. *Nature Reviews Drug Discovery*, **11**, 267–268.

11. www.ifpma.org/healthpartnerships

12. Smith, J. (2010) America's most generous companies. Forbes.com, October 28.

13. Herper, M. (2012) The Gardisil problem: How the U.S. lost faith in a promising vaccine. *Forbes*, April 23.

CHAPTER 6

FINAL THOUGHTS

FOR OVER a decade, the pharmaceutical industry has been battered by a host of issues: illegal marketing of drugs, safety of new medicines, lack of transparency, and so on. But perhaps the harshest criticisms have been around the very heart of the business: the discovery, development, and approval of new medicines.

The declining number of FDA approvals has only supported the criticism. From 1990 to 1999, the FDA approved an average of 31 drugs per year. In the next 10 years, this number dropped to 24. The outlook worsened in 2010, which saw only 21 FDA approvals. Thus, a 2011 front page story in the *Wall Street Journal* was eye opening: "Drug Makers Refill Parched Pipelines."[1]

Huh? Were authors Rockoff and Winslow delusional?

Not quite. The data they present in the article are certainly encouraging. The story has its roots in testimony by FDA Drug Division Director Dr. Janet Woodcock, who told Congress that the FDA has approved a number of innovative medicines that "work differently or better than existing drugs or tackle ailments lacking good treatments." She went on to say: "We're seeing a lot of innovation, much more than in recent memory." Dr. Woodcock is talking about the same pharmaceutical industry R&D engine that impatient critics have dismissed as broken. According to a graph in the *Wall Street Journal* article, Rockoff and Winslow predict that drug approvals in the 2010–2019 period will exceed anything that the industry has ever produced (Figure 6.1). If this is true, what blessed news for patients around the world.

Unfortunately, the *Wall Street Journal* authors' explanation of this productivity jump is flawed. To quote:

> Today's new drug output appears to mark the beginnings of a payoff from a research reorientation the industry began undertaking several years ago.

Actually, the productivity surge, if it follows the projected path, cannot be traced to any reorientation. It is the result of research done in biotech and pharmaceutical laboratories in the 1990s.

To support their argument, Rockoff and Winslow list compounds in a table called, "Novel Drugs Recently Approved and in the Pipeline" (Table 6.1). The very first compound in the table, Benlysta (for treating Lupus), was discovered by Human Genome Sciences (HGS) and jointly developed with GlaxoSmithKline (GSK). HGS started the discovery program that led to Benlysta in 1996, started clinical trials with

Devalued and Distrusted: Can the Pharmaceutical Industry Restore Its Broken Image?
First Edition. John L. LaMattina.
© 2013 John Wiley & Sons, Inc. Published 2013 by John Wiley & Sons, Inc.

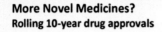

More Novel Medicines?
Rolling 10-year drug approvals

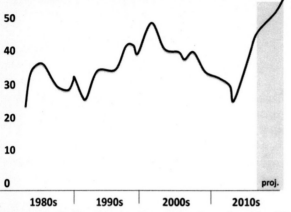

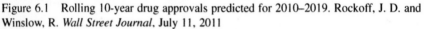

Figure 6.1   Rolling 10-year drug approvals predicted for 2010–2019. Rockoff, J. D. and Winslow, R. *Wall Street Journal*, July 11, 2011

this drug in 2001, and finally got approval this year. The discovery program that uncovered tofactinib, Pfizer's breakthrough drug for rheumatoid arthritis, started in 1993, and the New Drug Application (NDA) for this drug was filed in 2011. This drug was approved in late 2012; the R&D took 19 years! While I cannot attest to each of the compounds enumerated by the *Wall Street Journal*, my guess is that, based on the length of time it takes to get a drug approved, most or all of them had their roots in programs that commenced in the late 1990s.

If this is the case, what led to the surge that we are now seeing? As was discussed in Chapter 2, roughly 10 years ago two seismic changes greatly impacted the industry's productivity. The first involved the question: "What value does this medicine bring over existing therapies?" This question was asked not just by payers but also by regulatory agencies. Before 2000, studies to find answers were generally done after a drug was approved—so-called Phase 4 studies. However, in order to achieve a reasonable price for a new medicine, these studies now must be included in the initial filing for approval. The importance and costs of such studies cannot be underestimated. In many cases, studies that measure the performance of a new drug in a real-life situation are necessary. For example, it was no longer sufficient to show that a new medicine just lowered bad cholesterol. It also had to be proven that it could also reduce heart attacks and strokes. These studies alone add 3–5 years to the clinical development program and literally hundreds of millions of dollars in costs. Since many clinical programs were already underway in the mid-2000s, they had to be adjusted and so the programs took longer than originally planned.

In addition, by the mid-2000s, far more safety data were being required by the FDA than ever before. For example, the older NSAID pain relievers like naproxen

TABLE 6.1   Novel Drugs Recently Approved and in the Pipeline

Drug	Target	Timing	Maker	Estimated peak yearly sales
Benlysta	Lupus	2011	GlaxoSmithKline Human Genome Sciences	$3.2 billion
Yervoy	Metastatin melanoma	2011	Bristol-Myers Squibb	$2.8
Victrelis	Hepatitis C	2011	Merck	$1.2
Xarelto	Blood clots	2011	Bayer, Johnson & Johnson	$4.3
crizotinib	Lung cancer	2011	Pfizer	$2.0
vemurafenib	Metastatic melanoma	2011	Roche, Daichi Sankyo	$1.0
tofacitinib	Rheumatoid arthritis	2012	Pfizer	$2.2
bardoxolone	Chronic kidney disease	2013	Abbott Labs	$1.1
mericitabine	Hepatitis C	2014	Roche, Pharmasset	$1.3
Anti-BAFF antibody	Rheumatoid arthritis Lupus	2015	Eli Lilly	$1.2
darapladib	Atherosclerosis	2016	GlaxoSmithKline	$3.8

Source: Rockoff, J. D. and Winslow, R. Wall Street Journal, July 11, 2011

and ibuprofen had little long-term patient exposure at the time their respective NDAs were approved. To put it into perspective, all that was required in the past was a study showing safety in patients exposed to a drug for 90 days. Now, the FDA won't approve any pain reliever without patients being treated for at least a year (and more likely three years) with a study that also measures impact on overall health outcomes.

Are these important changes? Absolutely, because any drug that can get through these hurdles will, as Dr. Woodcock said, "work better than existing treatments." However, companies had to adjust to these changes in the last decade, and this drove up costs, caused development programs to take longer, and also resulted in more late-stage failures because compounds that were safe and effective might not have been as effective as existing, cheaper treatments and thus not commercially viable.

Given all of this information, I hope that the industry has adapted development programs and that timelines have been adjusted, and we can now expect a steady stream of new medicines. Yet, three issues temper my optimism. First, the industry has seen such an increase in productivity before. This happened in the mid-1990s when Congress enacted the Prescription Drug User Fee Act (PDUFA). The FDA was grossly understaffed in the 1990s and, as a result, many compounds languished awaiting approval while they were being approved in Europe, oftentimes years before they would be available in the United States. Congress was outraged by this. When the FDA showed data that indicated how understaffed it was, Congress's solution was the PDUFA, which essentially charged a company a fee when it filed an NDA and then used the revenues generated to hire more FDA reviewers. This action resulted in the FDA being able to review dossiers more rapidly, which led to the removal of the logjam and more medicines being approved than any time before or since. Has the adaptation of the industry to the new NDA expectations led to a similar increase of compounds being approved? We won't know for 5 years.

Second, the consolidation of the industry in the last 15 years, as discussed in Chapter 2, has raised havoc with R&D organizations. It is my firm belief that mergers are particularly difficult for R&D because starting/stopping research programs takes time. I once heard a Nobel Prize winner in Medicine say that it takes at least 3 years for a professor, who leaves one university for another, to get his laboratory back running at full speed. My experience is that it is a lot quicker in industry, but it still takes time.

Finally, the cuts that have been made across the board in R&D in many companies will have an impact going forward. This effect will not be seen in the short term. As was detailed above, discovery–development programs for successful new medicines take over a decade. However, the turmoil in R&D of the recent past (mergers, reorganizations, site closures, new business models, etc.) will be felt in the next decade.

However, the R&D cuts being made across the pharmaceutical industry are most concerning to me.

Comments by Kenneth Frazier, Merck's CEO, should have served as an inspiration for those focused on the promise of new medicines and perhaps acted as a beacon for other leaders in the pharmaceutical field. Frazier said that Merck will not focus on cuts, but rather focus on investing in drug development to drive growth.[2] He said that he wanted to reinvest some of the savings realized from the acquisition of Schering-Plough into advancing Merck's last-stage pipeline. Frazier continued on this theme at his company's annual shareholder meeting in May, where he shared his vision of success by using Apple, IBM, and Starbucks as examples. These companies succeeded through innovation, and he said that innovation and research were key to his vision for Merck.[3]

Wall Street analysts, however, weren't happy. Les Funtleyder, a manager of the Miller Taback Health Care Transformation Fund, provided a typical response: "Merck could end up wasting billions of dollars probing compounds that don't pan out." Analysts contrasted Merck's approach to that of Pfizer, which was cutting its R&D budget drastically. The result: Pfizer's stock price advanced while Merck's took a hit.

The Merck situation is not unique. John Lechleiter, Lilly's CEO, has also committed to strong R&D investments for his company, saying that "It would be a mistake for us to disinvest in any significant way in R&D." He also realizes that such a strategy is frowned upon by Wall Street: "I never thought that I would live to see this, but investors are actually thinking to cut R&D—that's the hot topic of the day. This is kind of nuts, but this is what is being talked about."[4] Analysts have been negative on Lechleiter's stance.

Now, I am all for monitoring R&D budgets to maximize the returns from these investments. And I am all for accountability—asking the R&D organization to deliver new candidates to the pipeline, having formal goals with rigorous deadlines, and running clinical trials as expeditiously as possible while keeping a close eye on costs. But for Wall Street to reward a company for lowering R&D spending and attack those that want to commit to R&D is absurd. Like it or not, R&D IS the engine that powers a pharmaceutical company. It is also a high-risk endeavor. Furthermore, given all of the hurdles that now exist especially with regard to ensuring safety and having sufficient novelty to justify pricing, R&D is more expensive than ever. But, if you want to succeed, you have to invest—substantially. There are no short cuts.

But the decision on how much to invest in R&D, the lifeblood of a company, is the responsibility of the CEO and the company's Board of Directors. For this to be influenced by Wall Street Analysts is, to quote Lechleiter, "Nuts."

Let's say that the industry really has turned around its productivity issues. Perhaps the recent surge in new product approvals is not a mirage and is sustained. Maybe Dr. Woodcock's words about seeing more innovation than at any time in recent memory will be echoed for years to come. Will that change public opinion? Will future audiences of *The Dr. Oz Show* welcome representatives of the drug industry as valued contributors to the nation's health care?

Probably not. The industry still has a big hill to climb to restore public confidence in the value of new medicines in terms of both their healing properties and their risk. The industry doesn't invent diseases—it tries to prevent or alleviate them. The industry diligently tries to understand the risks inherent in their products and works with patients and physicians to put side-effects in their proper context. Industry representatives meet with physicians routinely to discuss new medicines and to share data that can help benefit their patients—sessions designed for mutual learning and not meant for undue influence. The pharmaceutical industry was founded on these principles. Have they strayed? Sure, but that doesn't mean that the vast majority of colleagues working at these companies aren't committed, skilled people whose major career driver is to use their special talents to bring forward new medicines to benefit millions of people around the world.

Will the mother discussed in the opening of this book who tragically lost her daughter ever believe this? I am not sure. But, if the pharmaceutical industry ever were able to regain its "most admired" status, such suspicions and animosity would be rare. Patients would accept the fact that any drug carries risks, but that the innovative company that brought this new drug to market did everything it could to ensure its efficacy and safe use.

Ironically, we are in an era where the role of drug companies couldn't be more important. Every day, new discoveries are made to elucidate the cause and etiology of many diseases. However, a vibrant and robust pharmaceutical industry is needed to build on these basic understandings and turn them into new medicines. A successful pharmaceutical industry benefits all of us. Hopefully, the recommendations made in this book will be useful in moving the industry to a better place.

REFERENCES

1. Rockoff, J. D., Winslow, R. (2011) Drug Makers Refill Parched Pipelines. *Wall Street Journal*, July 11.
2. Rockoff, J. D. (2011) Pfizer, Merck Take Different R&D Tacks. *Wall Street Journal*, February 4.
3. Todd, S. (2011) Merck CEO Frazier stresses innovation will be key to drug makers' success. *The Star Ledger*, May 24.
4. Pierson, R. (2011) Lilly CEO defends R&D spending. Reuters, June 30.

INDEX

Note: Page numbers in *italics* indicate figures; tables are noted with *t*.

Devalued and Distrusted: Can the Pharmaceutical Industry Restore Its Broken Image?
First Edition. John L. LaMattina.
© 2013 John Wiley & Sons, Inc. Published 2013 by John Wiley & Sons, Inc.

CPSIA information can be obtained at www.ICGtesting.com
Printed in the USA
BVOW01n0333070214

344171BV00001B/1/P

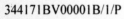